Contents

Oscillations and waves

Quantum physics

Our universe

Introduction

This series has been written by Principal Examiners and others involved directly with the development of the Edexcel Advanced Subsidiary (AS) and Advanced (A) GCE Physics specifications.

Waves and Our Universe is one of four books in the Nelson Advanced Science (NAS) series developed by updating and reorganising the material from the Nelson Advanced Modular Science (AMS) books to align with the requirements of the Edexcel specifications from September 2000. The books will also be useful for other AS and Advanced courses.

Waves and Our Universe provides coverage of Unit 4 of the Edexcel specification.

Other resources in this series

NAS Teachers' Guide for AS and A Physics provides a proposed teaching scheme together with practical support and answers to all the practice and assessment questions provided in *Mechanics and Radioactivity; Electricity and Thermal Physics; Waves and Our Universe;* and *Fields, Forces and Synthesis.*

NAS Physics Experiment Sheets 2nd edition by Adrian Watt provides a bank of practical experiments that align with the NAS Physics series. They give step-by-step instructions for each practical provided and include notes for teachers and technicians.

NAS Make the Grade in AS and A Physics is a Revision Guide for students. It has been written to be used in conjunction with the other books in this series. It helps students to develop strategies for learning and revision, to check their knowledge and understanding and to practise the skills required for tackling assessment questions.

Features used in this book

The Nelson Advanced Science series contains particular features to help you understand and learn the information provided in the books, and to help you to apply the information to your studies.

About the authors

Mark Ellse is Principal of Chase Academy in Cannock, Staffordshire, and former Principal Examiner for Edexcel.

Chris Honeywill is a Reviser and Acting Principal Examiner for Edexcel and former Head of Physics at Farnborough Sixth Form College.

Oscillations and waves

Sea waves can have enormous energy

Going round in circles

Figure. 1.1 *Circular motion can be fun!*

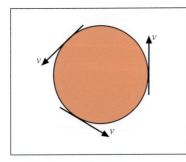

Figure 1.2 *In uniform circular motion the speed of a body is constant but its direction continually changes*

Uniform circular motion

Many things travel in circular paths. If you sit on a roundabout, follow a circular road, or just stand on the moving surface of the Earth, you are performing circular motion.

The Moon's movement around the Earth is approximately circular motion, as are the movements of the planets around the Sun. The most simple type of circular motion to study is one in which a body is travelling at constant speed – for example, if you cycle along a circular path at constant speed, or if you sit on a roundabout going at a constant speed (Figure 1.1). This type of motion is called **uniform circular motion**.

Changing direction

The speed of a body in uniform circular motion is constant. But, as Figure 1.2 shows, the body's *direction* of motion is continually changing.

The body's velocity is a vector quantity that depends on both speed and direction. The direction is continually changing, so the velocity of the body is continually changing. There is no change in the *magnitude* of the velocity (the speed), only the *direction* of the velocity.

Centripetal acceleration

If the velocity of a body is changing, it must be accelerating.

Figure 1.3 shows how to find the direction of the acceleration:

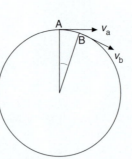

The vector triangle shows *v*, the change in velocity. The change is towards the centre of the circle

As the body moves from A to B, its velocity changes from v_a to v_b

Figure 1.3 *Finding the direction of the acceleration*

Since the change of velocity is toward the centre of the circle, the acceleration is also toward the centre (Figure 1.4). This is called centripetal (centre-seeking) acceleration. A body performing uniform circular motion has centripetal acceleration.

Centripetal acceleration is at right angles to the velocity, toward the centre of the circle. It does not cause an increase or decrease in the speed of a body, only a change in its direction.

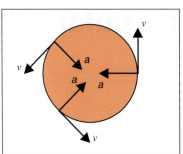

Figure 1.4   In circular motion, the acceleration is toward the centre of the circle.

Centripetal force

Figure 1.5 shows a motorcyclist going round a roundabout. The motorcyclist is performing circular motion – he is accelerating toward the centre of the roundabout. What is causing the motorcyclist to accelerate?

Figure 1.5   Friction from the ground provides the centripetal force

You know that there must be a force on the motorcyclist, causing the acceleration. Since the motorcyclist is accelerating toward the centre, the force must be toward the centre. The friction between the motorcyclist and the road provides the force. The centre-seeking force that causes a body to do circular motion is called the centripetal force.

With circular motion, there is both force and motion, but the kinetic energy of the body does not change. So, the resultant force is not doing any work on the body. The force and acceleration are toward the centre (Figures 1.4 and 1.6), but the velocity is at right angles to these. The distance moved in the direction of the force is zero.

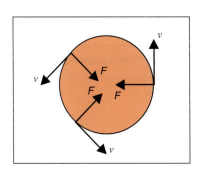

Figure 1.6   In circular motion, forces act towards the centre of the circle

Circular motion calculations

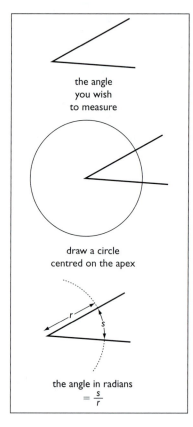

Figure 2.1   Measuring an angle in radians

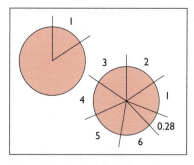

Figure 2.2   One radian is equal to 57°, and there are 2π(6.28) radians in a circle

Angles in circular motion

For circular motion, you need to be able to measure angles in **radians**. To find the size of an angle in radians, draw a circle centred on the apex and divide the length of the arc by the radius (Figure 2.1). One radian is just over 57°. There are $2\pi = 6.28$ radians in a whole circle (Figure 2.2).

Calculating centripetal force

The centripetal acceleration of a body performing uniform circular motion depends on its speed, v and the radius of the path, r:

$$\text{centripetal acceleration} = \frac{v^2}{r}$$

Figure 2.3 shows how you can use geometry to prove this relationship.

Since:

$$\text{force} = \text{mass} \times \text{acceleration}$$

$$\text{centripetal force} = \frac{mv^2}{r} \text{ where } m \text{ is the mass of the body.}$$

The motorcycle and rider in Figure 1.5 have a mass of 280 kg. They are travelling at a speed of 32 m s^{-1} around a circle of radius 130 m.

$$a = \frac{v^2}{r} = \frac{(32 \text{ m s}^{-1})^2}{130 \text{ m}} = 7.9 \text{ m s}^{-2}$$

$$F = ma = 280 \text{ kg} \times 7.9 \text{ m s}^{-2} = 2200 \text{ N}$$

Period and frequency

The **period** T of a body doing uniform circular motion is the time it takes to complete one revolution. Period is usually measured in seconds. The **frequency** of rotation is the number of rotations per second.

Frequency = 1/period; $f = 1/T$. The unit of frequency is s^{-1}, called the hertz (Hz).

Angular speed

The **angular speed** ω of a body is the central angle through which it turns in a second (Figure 2.5):

$$\omega = \theta/t$$

It takes a time T for a body to complete one revolution about the centre. During this time the body rotates through an angle of 2π radians about the centre. So

$$\omega = \frac{\theta}{t} = \frac{2\pi}{T}$$

Since $T = \dfrac{1}{f}$ $\qquad \omega = \dfrac{2\pi}{T} = \dfrac{2\pi}{\left(\frac{1}{f}\right)} = 2\pi f$

WORKED EXAMPLE

The rate of rotation of a compact disc varies, but a typical value is 78 revolutions per minute. So

$$\text{period} = \frac{60\,\text{s}}{78} = 0.77\,\text{s} \quad \text{and} \quad \text{frequency} = \frac{1}{T} = \frac{1}{0.77\,\text{s}} = 1.3\,\text{Hz}$$

angular speed

$$\omega = 2\pi/T = 2\pi\,\text{rad}/0.77\,\text{s} = 8.2\,\text{rad s}^{-1}$$

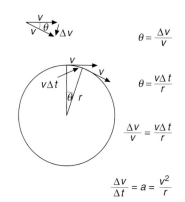

$$\theta = \frac{\Delta v}{v}$$

$$\theta = \frac{v\Delta t}{r}$$

$$\frac{\Delta v}{v} = \frac{v\Delta t}{r}$$

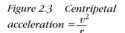
$$\frac{\Delta v}{\Delta t} = a = \frac{v^2}{r}$$

Figure 2.3   Centripetal acceleration $= \dfrac{v^2}{r}$

Calculating force and acceleration from ω

When a body is doing uniform circular motion, it travels right round the circumference, a distance $2\pi r$, in a time T. You know that speed = distance/time, so

$$v = \frac{2\pi r}{T}$$

But $T = 2\pi/\omega$, so

$$v = \frac{2\pi r}{(2\pi/\omega)}$$

$$v = r\omega$$

You know that centripetal acceleration $a = \dfrac{v^2}{r}$. So

$$a = \frac{(r\omega)^2}{r} = \frac{r^2\omega^2}{r} = r\omega^2$$

and

$$\text{centripetal force} = ma = mr\omega^2$$

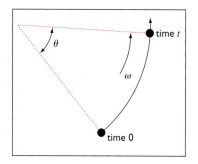

Figure 2.4   Angular speed $\omega = \theta/t$

Investigating centripetal force

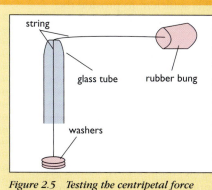

Figure 2.5   Testing the centripetal force equation

- Measure the mass of the bung. Put a mark on the string 2 m from the bung; thread the string through the glass tube and put a weight of 0.5 N on the other end.
- Hold the glass tube and swing the bung around in a horizontal circle (Figure 2.5) in such a way as to achieve a constant radius of 2 m. If there is little friction between the string and the glass tube you can assume the weight is equal to the centripetal force.
- Find the time for one rotation and use this to calculate the speed of the bung.
- Compare the centripetal force given by mv^2/r with the force of 0.5 N provided by the weight.
- Repeat with different weights and different radii.

3 Circular motion under gravity

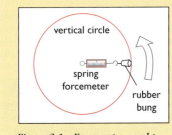

Figure 3.1 Forcemeter used to monitor the force on a bung

- Observe the forcemeter as you whirl a bung in a vertical circle (Figure 3.1).
- Whirl the bung quickly, then as slowly as you can whilst still keeping it moving in a circle. You will not be able to make it move at constant speed in a circular path, but do your best.
- When the bung is moving slowly, notice how the tension varies around the circle.

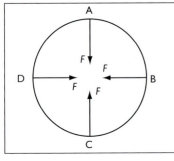

Figure 3.2 Required centripetal force has constant magnitude

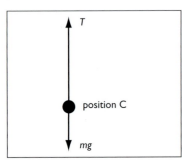

Figure 3.3 Resultant provides upward centripetal force

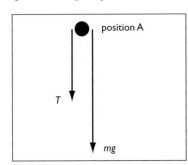

Figure 3.4 Resultant provides downward centripetal force of same magnitude

Motion in a vertical circle

When a body is moving at constant speed in a vertical circle, the centripetal acceleration is the same size at all positions around the circle and always directed towards the centre. So the resultant force on the body, the centripetal force, also has the same size and is directed towards the centre of the circle (Figure 3.2).

If you swing a bung round vertically, the centripetal force is provided by the resultant of two forces – the force of the string and the force of gravity.

At point C in Figure 3.2 the string is pulling towards the centre, and the weight is pulling away from the centre (Figure 3.3). The string has to support the weight of the body as well as providing the upward centripetal force acting on it. The tension is at its greatest at this point.

At point A the weight of the body is providing some of the downward centripetal force (Figure 3.4). The string has only to provide the remainder. The tension is at its smallest at this point. As you reduce the speed of rotation, there comes a time when the weight alone is sufficient to provide the centripetal force. At that speed, the tension drops to zero at the top.

Loop-the-loop rides

Look at the ride shown in Figure 3.5. At the bottom of the loop, the upward contact force from the track equals the weight of the vehicle plus the required centripetal force. At the top, the track and the weight together provide the centripetal force. If the speed is slow enough at the top, the weight alone is exactly the required centripetal force and the track no longer exerts any force on the vehicle and its passengers. Indeed, you could remove a small section of track at the top, and the vehicle would continue unaffected, falling freely, but still following the circular path.

Weightlessness and apparent weightlessness

Figure 22.3 in *Mechanics and Radioactivity* showed how gravity produces tensions and compressions in parts of your body. These give you the experience of having a weight. If you were a long way from all other masses, you would be **weightless**. You would feel weightless because you would no longer feel the tensions and compressions that weight produces.

If you fall freely, for instance if you jump off a high diving board (Figure 3.6), the downwards gravitational forces are still there but there are no supporting forces. All parts of your body accelerate downwards at a rate equal to *g*. The lack of upwards forces means that the normal tensions and compressions are not there, so it feels as if you are weightless. This is **apparent weightlessness**.

If you go over a loop-the-loop ride where the only force acting on you at the top is weight, you would feel weightless as you fall freely towards the Earth.

An astronaut in an orbiting spacecraft (Figure 3.7) is falling freely towards the Earth. His weight is the centripetal force for his circular motion, and there are no supporting forces. Although he has a weight, he feels weightless.

The conical pendulum

Figure 3.8 shows a conical pendulum. The bob is moving at a constant speed in a horizontal circle. There are two forces acting on the bob: its weight and the tension in the string (Figure 3.9). The circle is horizontal, so there is no vertical motion; the bob is in vertical equilibrium. Therefore the vertical component of the tension equals the weight of the bob:

$$T \cos \theta = mg$$

The circular motion requires a resultant horizontal force towards the centre. The horizontal component of the tension provides this:

$$T \sin \theta = \frac{mv^2}{r}$$

Combining these two expressions gives

$$\frac{T \sin \theta}{T \cos \theta} = \frac{\frac{mv^2}{r}}{mg} = \frac{mv^2}{rmg} = \frac{v^2}{rg}$$

$$\tan \theta = \frac{v^2}{rg}$$

If a conical pendulum is swinging in a horizontal circle of radius 1.5 m at an angle of 10° to the vertical, numerically

$$v^2 = rg \tan \theta = 1.5 \times 9.8 \times \tan(10°)$$

$$v^2 = 2.59 \text{ m}^2 \text{ s}^{-2}$$

$$v = 1.6 \text{ m s}^{-1}$$

Figure 3.5   People enjoying themselves on a loop-the-loop ride!

Figure 3.6   If you jump off a high diving board, you experience apparent weightlessness

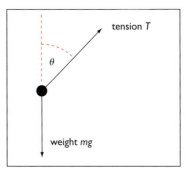

Figure 3.7   An astronaut 'floating' inside a spacecraft

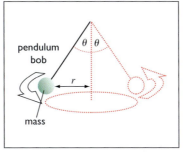

pendulum bob

r

mass

Figure 3.8   A conical pendulum

tension *T*

θ

weight *mg*

Figure 3.9   Free-body force diagram for the bob

4 # Periodic motion

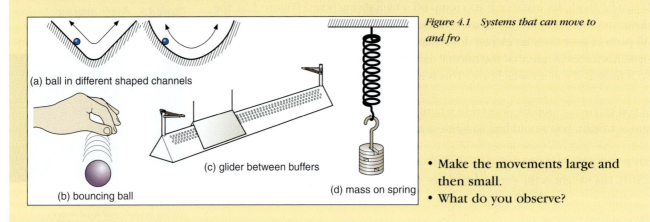

Observing repeating motions

• Observe the systems shown in Figure 4.1. Set them going quickly, and then slowly.

(a) ball in different shaped channels

(b) bouncing ball

(c) glider between buffers

(d) mass on spring

Figure 4.1   Systems that can move to and fro

• **Make the movements large and then small.**
• **What do you observe?**

Oscillations

Many systems have repeated movements. But the motion of some systems is regular, and that of other systems is irregular. The swings of a particular pendulum all take the same time, however fast the pendulum is swinging. But the to-and-fro motion of a glider bouncing between the ends of an air track depends on the speed of the glider.

The motion of your legs as you walk steadily, the vibrations of a mechanical road digger, the movements of piano strings – these are all repeating motions that are regular. Movements like these are particularly interesting and useful. Since they are regular, you can use them for clocks. Regular movements like these are called **oscillations**. The bodies that oscillate move to and fro either side of their **equilibrium position**, the position in which the resultant force on them is zero, and in which they would be at rest if not oscillating.

At any instant during an oscillation, you can record the position of the body – its **displacement** – from its equilibrium position. When a body is oscillating, it goes either side of its equilibrium position. You can use positive values to indicate displacements on one side, and negative values to indicate displacements in the opposite direction, as Figure 4.2 shows. You can produce a **time trace**, which is a graph to show how the displacement of an oscillation varies with time.

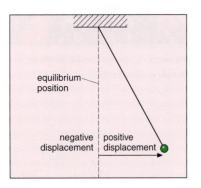

Figure 4.2   The displacement of this pendulum is positive

equilibrium position

negative displacement | positive displacement

8

Producing time traces

- The pendulum in Figure 4.3 swings from a potentiometer. The voltage from the circuit depends on the displacement. Set the pendulum swinging and use the oscilloscope to observe how the displacement varies with time. Sketch a graph for the motion.
- Set up a mass oscillating on the end of a spring as in Figure 4.4. Place a motion sensor connected to a datalogger (Figure 4.5) below the oscillating mass. Use the datalogger to display a displacement–time graph for the oscillating mass.
- Compare the time traces for each of the above oscillations. What things do they have in common?

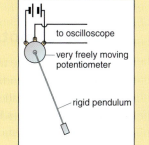

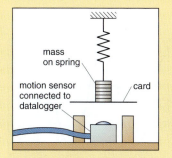

Figure 4.3   Producing a time trace for a pendulum

Figure 4.4   Using a motion sensor to produce a time trace

Amplitude, period and frequency

As Figure 4.6 shows, the **amplitude** of an oscillation is the maximum displacement from the equilibrium position. For each of the oscillating systems in the experiment, the amplitudes of oscillation decrease as the oscillations gradually die away. A complete movement of an oscillating system – from its equilibrium position, to a displacement in one direction, back to the equilibrium position, to a displacement in the reverse direction and finally back to its equilibrium position again – is called one **cycle**. The oscillating systems in the experiments above are regularly repeating. The time for one cycle, called the **period**, is constant. The period does not change when the amplitude decreases as the oscillations die away.

Figure 4.5   A datalogger

It is convenient to use the period to describe the rate of slow oscillations. The period of the pendulum of a grandfather clock is 2 s. But many oscillations are *very* quick, and if the period is short, it is more usual to describe the rate of oscillation by the number of complete oscillations per second – the **frequency**, measured in hertz (Hz). The frequency of the sounds you hear varies from about 30 Hz up to about 20 000 Hz – that is, from 30 to 20 000 oscillations per second (or 30 s^{-1} to 20 000 s^{-1}).

As with circular motion you can calculate the period from the frequency:

$$\text{period} = \frac{1}{\text{frequency}} \quad \text{or} \quad T = \frac{1}{f}$$

For the top of the audible frequency range,

$$T = \frac{1}{f} = \frac{1}{20\ 000\ \text{Hz}} = \frac{1}{20\ 000\ s^{-1}} = 50\ \mu s$$

Figure 4.6   As the oscillation dies away, the amplitude decreases

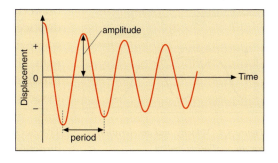

5 Simple harmonic motion

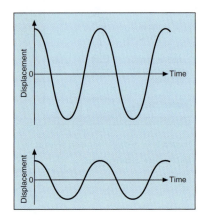

Figure 5.1   The period of oscillation is independent of amplitude

Sinusoidal motion

Figure 5.1 shows time traces for a trolley tethered by springs. The period of these oscillations is independent of amplitude – it remains the same whether the trolley is making large oscillations, or small ones. Motion like this is called **simple harmonic motion (s.h.m.)**. Displacement–time graphs for simple harmonic motion all have a characteristic shape – called **sinusoidal**, the shape of a mathematical sine or cosine graph.

Velocity and displacement

You can deduce a velocity–time graph for simple harmonic motion from the gradient of a displacement–time graph. As Figure 5.2 shows, the gradient of the displacement–time graph is zero when the displacement is a maximum. The velocity is zero at these points. For a trolley between two springs, the velocity is zero for a moment at either end of the motion, where the displacement is maximum positive, or maximum negative.

The velocity is maximum positive, or maximum negative, when the displacement is zero – when the trolley is in the centre of the motion, moving fastest either to left or to right.

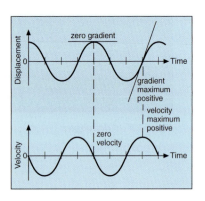

Figure 5.2   Displacement and velocity for an oscillating trolley

Acceleration and velocity

You can deduce the acceleration–time graph from the velocity–time graph. As Figure 5.3 shows, when the velocity is maximum, the gradient of the velocity–time graph is zero, and therefore the acceleration is zero. The acceleration is zero in the middle of the motion, when the velocity is maximum positive or maximum negative.

The acceleration is maximum, positive or negative, when the velocity is zero – when the trolley is momentarily stationary at the extremes of its motion.

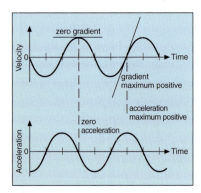

Figure 5.3   Velocity and acceleration for an oscillating trolley

Acceleration and displacement

Figure 5.4 puts Figures 5.2 and 5.3 together and shows the relationship between acceleration and displacement. The acceleration is zero when the displacement is zero – in the middle of the motion. The acceleration is large where the displacement is large, but to the left when the displacement is to the right. The acceleration is maximum to the left when the displacement is maximum to the right – and vice versa. The acceleration is always directed towards the equilibrium position.

The acceleration is proportional to the displacement, but in the opposite direction to the displacement. Mathematically:

$$\text{acceleration} \propto -\text{displacement} \qquad \text{or} \qquad a \propto -x$$

Figure 5.5 shows a graph of this relationship. This relationship between acceleration and displacement produces simple harmonic motion. Often we write the s.h.m. equation as:

$$a = -\omega^2 x$$

(ω is the Greek letter *omega*.) ω^2 is always positive, so the acceleration and the displacement are always of opposite sign. Just as with circular motion, ω is called the angular speed and is related to the period by the equation

$$T = \frac{2\pi}{\omega}$$

Force and displacement

You know that force is proportional to acceleration and, since acceleration is proportional to negative displacement, force must be proportional to negative displacement:

$$F \propto -x$$

You can define simple harmonic motion as the motion that occurs when a body has an acceleration (or a force) that is directly proportional to its displacement from the equilibrium position and always directed towards that central position.

For a trolley tethered between two springs, the force F is proportional to the displacement x. The force is always in the opposite direction to the displacement, so we can write:

$$F = -kx$$

where k is the spring constant. Since $F = ma$, where m is the mass and a is the acceleration, we can write

$$ma = -kx \quad \text{and so} \quad a = -\left(\frac{k}{m}\right)x$$

If you compare this equation with $a = -\omega^2 x$, you can see that

$$\frac{k}{m} = \omega^2 \quad \text{and so} \quad \omega = \sqrt{\frac{k}{m}}$$

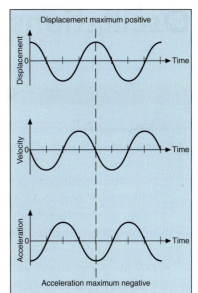

Figure 5.4   Acceleration and displacement for an oscillating trolley

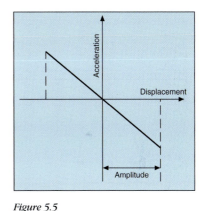

Figure 5.5
Acceleration = $-\omega^2$ displacement

Measuring ω for an oscillating trolley

- Measure the mass m of a trolley and then tether it between two springs as shown in Figure 5.6.

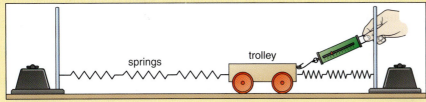

Figure 5.6   Measuring the spring constant for a tethered trolley

- Use a spring forcemeter to measure the force needed to displace the trolley a measured distance from equilibrium.
- Calculate the spring constant k. From k/m calculate ω.
- Set the trolley oscillating and measure the period. Check the relationship between the period and ω.

Oscillations and circular motion

Circular paths and linear oscillators

- Tether a trolley between two springs and add mass to it to give it a period of about 2 s.
- Place the rotating disc centrally behind the trolley (Figure 6.1). Adjust the motor so that the time the peg takes to complete one orbit is the same as the period of the trolley.
- Pull the trolley to the left and release it as the peg gets furthest to the left.
- Compare the two motions.

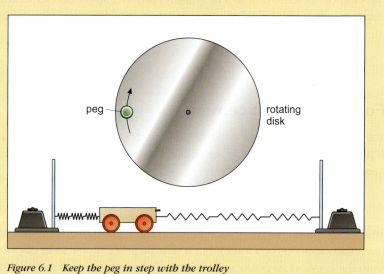

Figure 6.1   Keep the peg in step with the trolley

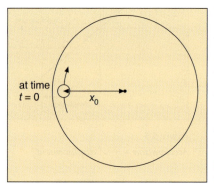

Figure 6.2   At time t = 0, the peg is a maximum distance to the left

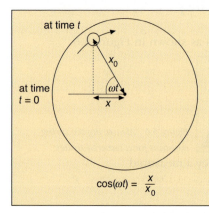

Figure 6.3   After a time t, the disc has rotated through an angle ωt

The shadow of a circular path

The peg on the disc in Figure 6.1 moves from side to side in step with the oscillating trolley. The two motions remain in phase (in step) with each other. The side-to-side motion of the trolley with its speed constantly increasing and decreasing is identical to the side-to-side motion of an object moving around a circular path at a constant speed. The angular speed ω for simple harmonic motion (s.h.m.) is the angular speed of circular motion that keeps in step with that s.h.m. The period T of the circular motion is the same as the period of the s.h.m. of the trolley, and is related to the angular velocity by the equation

$$T = \frac{2\pi}{\omega}$$

Figure 6.2 shows the maximum displacement of the peg to the left, a distance x_0 from the centre of the disc. If the angular speed of the disc is ω, after a time t the disc has rotated through an angle ωt, as Figure 6.3 shows. The horizontal displacement is now x. So from the triangle in Figure 6.3:

$$\cos(\omega t) = \frac{x}{x_0} \qquad \text{or} \qquad x = x_0 \cos(\omega t)$$

Figure 6.4 shows a graph of displacement x against time t. As you would expect, it is just like the time trace for s.h.m., since both vary sinusoidally with time. The maximum and minimum values of $\cos(\omega t)$ are $+1$ and -1; x can vary between x_0 and $-x_0$. So you can see that the amplitude of the motion is x_0.

The mathematics of s.h.m.

If you can differentiate, you can use this to find the equations for velocity v and acceleration a from the equation for the displacement x. Since

$$x = x_0 \cos(\omega t)$$

then

$$v = \frac{dx}{dt} = -\omega x_0 \sin(\omega t)$$

and

$$a = \frac{dv}{dt} = -\omega^2 x_0 \cos(\omega t)$$

These are the equations of the three graphs shown in Figure 5.4 in the last chapter. Since

$$a = -\omega^2 x_0 \cos(\omega t)$$

and

$$x = x_0 \cos(\omega t)$$

then

$$a = -\omega^2 x$$

This is the defining equation for s.h.m.
Since the values of both sine and cosine vary between $+1$ and -1, the maximum velocity is ωx_0 and the maximum acceleration is $\omega^2 x_0$.

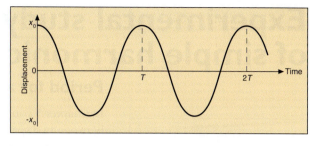

Figure 6.4   Horizontal displacement of the peg

Phase angles

Imagine two systems that are rotating or oscillating exactly in step. The displacement–time graphs for both of these oscillations would look like Figure 6.4 (although the graphs might have different amplitudes). These two objects are like two friendly runners on a circular track. At all times they are at exactly the same point as each other. These two objects are said to be **in phase**.

Figure 6.5 shows graphs for two runners who wish to keep as far apart as possible. They remain opposite each other on the track by moving around it at the same speed. The time traces are the same as those of two oscillators that are half an oscillation apart. These are exactly **out of phase** with each other or **in antiphase**. The phase angle difference between them is π radians (a straight line across the centre of the circle).

A phase difference of $\pi/2$ radians is shown in Figure 6.6.

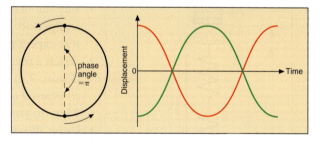

Figure 6.5   Moving in antiphase

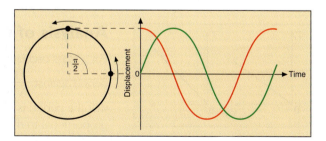

Figure 6.6   A phase difference of $\pi/2$ radians

7 Experimental study of simple harmonic motion

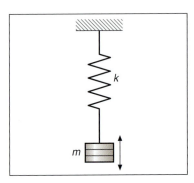

Figure 7.1   Vertical oscillation

Period for a mass–spring system

In Chapter 5 you learned that the angular speed ω for a trolley tethered between two springs depends on the mass of the trolley m and the spring constant of the spring system k:

$$\omega = \sqrt{\frac{k}{m}}$$

Since the period $T = 2\pi\backslash\omega$, then

$$T = 2\pi\sqrt{\frac{m}{k}}$$

You can investigate experimentally how the period for a mass oscillating vertically on the end of a spring depends on k and m.

Another mass–spring system

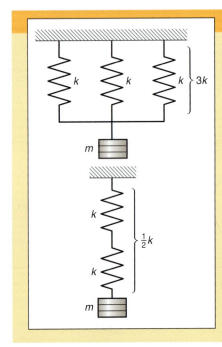

- Find the average spring constant of several identical springs. Use one of the springs and determine the period of oscillation of a known mass suspended from it (Figure 7.1). Time at least ten oscillations to increase the accuracy of your readings.
- Repeat your readings.
- Vary the mass and record a set of corresponding readings of mass m and period T. Plot a graph of T against $\sqrt{m}$. What relationship do you expect to find?
- Attach a mass of 500 g in turn to different spring combinations (Figure 7.2) and record the period T together with the spring constant k of the combination used. Plot a graph of T against $\frac{1}{\sqrt{k}}$. What do you expect to find?
- Using all your results, plot a graph of T against $\sqrt{\frac{m}{k}}$ and measure the gradient.

Figure 7.2   Producing different spring constants

Analysis of results

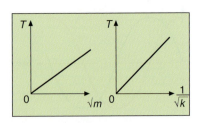

Figure 7.3   How mass and spring constant affect period

For a mass oscillating at the end of a spring, graphs of period against $\sqrt{m}$ and period against $\frac{1}{\sqrt{k}}$ are straight lines through the origin (Figure 7.3). This shows that the period is proportional to the square root of the mass and inversely proportional to the square root of the spring constant, i.e.,

$$T \propto \sqrt{\frac{m}{k}}$$

As you might expect, the constant of proportionality in this equation is 2π, so

$$T = 2\pi\sqrt{\frac{m}{k}}$$

just like the trolley oscillating between two springs.

A simple pendulum

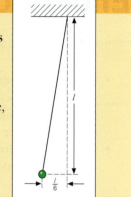

- The motion of a pendulum is simple harmonic only when the amplitude of swing is small – less than about a sixth of the length of the pendulum (Figure 7.4).
- For small-angle swings, the period might depend on (a) the mass of the bob and/or (b) the length of the string. Predict how each of these variables affects the period.
- Carry out an experiment to test your prediction. Change only one variable at a time, and use a graph to show how period depends on the other variable.

Figure 7.4   Pendulum oscillations are simple harmonic only for small-angle swings

Period equation

For small-angle swings, the period of a pendulum of length *l* is given by

$$T = 2\pi \sqrt{\frac{l}{g}}$$

This shows that the period depends only on the length *l* of the pendulum and *g*, the acceleration of gravity. A graph of *T* against $\sqrt{l}$ will be a straight line through the origin with gradient $2\pi/\sqrt{g}$. You can use this graph to calculate a value for the acceleration of gravity.

Use of a pendulum in time-keeping

Grandfather clocks like that in Figure 7.5 are often fitted with a *seconds* pendulum, which advances the mechanism twice an oscillation at 1 s intervals. The period of such a pendulum is therefore 2 s.

But what must be the length of this pendulum? Using $T = 2\pi \sqrt{\frac{l}{g}}$ gives

$$l = \left(\frac{T}{2\pi}\right)^2 g = \frac{T^2 g}{4\pi^2} = \frac{(2\ \text{s})^2 \times 9.8\ \text{m s}^{-2}}{4\pi^2} = 0.99\ \text{m}$$

This accounts for the height of such clocks, as their casings must be sufficiently long to house the pendulum. They are sometimes known as long-case clocks.

Figure 7.5   A long-case clock

Why is the period of s.h.m. constant?

For a long time, clocks used pendulums to keep time. Nowadays many clocks use a mass–spring system. Most modern clocks and wristwatches rely on a quartz crystal, where its own mass oscillates under its own springiness. In all these clocks, simple harmonic motion provides good time-keeping because the period is independent of the amplitude of oscillation. Think about why this is so.

If you double the amplitude of s.h.m., the period stays the same. The body has to travel twice the distance in the same time as before. Its average speed is doubled. This doubled speed is achieved in the same time, so the body has twice the average acceleration and twice the average resultant force acting on it, i.e., double the amplitude leads to double the maximum acceleration and double the force. The good time-keeping that s.h.m. provides is because force and acceleration are proportional to displacement.

Energy of an oscillator

Energy against time for an oscillator

- Use a motion sensor to produce a velocity–time graph for an oscillating trolley as shown in Figure 8.1.
- From the velocity–time graph, produce a kinetic energy–time graph.
- Predict the shape of the potential energy–time and total energy–time graphs.

Figure 8.1 Use the motion sensor to find how velocity varies with time

---

ADDITIONAL MATERIAL

Energy–time graphs

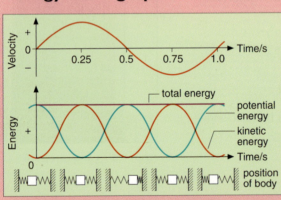

Figure 8.2 There is a constant interchange between kinetic and potential energy in an oscillator

The velocity–time graph in Figure 8.2 is for a body oscillating with period 1 s. The velocity–time graph is a sine wave. The velocity is zero at the beginning of the cycle, maximum positive 0.25 s later, zero half-way through the cycle at $t = 0.5$ s, maximum negative at 0.75 s and back at zero again at $t = 1$ s.

The red graph shows how kinetic energy varies. Kinetic energy $= \frac{1}{2}mv^2$: it is zero at $t = 0$, $t = 0.5$s and $t = 1.0$s, when the velocity is zero. Kinetic energy is maximum at $t = 0.25$ s, when the velocity is maximum positive, and also at $t = 0.75$ s, when the velocity is maximum negative.

The kinetic energy of the oscillator is a maximum when it is in the centre of its motion. A quarter of a cycle later, the kinetic energy is zero when the body is at one or other extreme of the motion.

The oscillator has potential energy stored in the compressed and stretched springs. The blue graph of Figure 8.2 shows how the potential energy varies with time. The potential energy is maximum at the extremes of the motion, and minimum in the middle.

The purple graph of Figure 8.2 is the total energy. It is the sum of the kinetic and potential energies. Energy is conserved. The graph is a horizontal straight line because the total energy of the oscillator remains constant.

Energy–displacement graphs

Figure 8.3 shows how the energy of an oscillator varies with the displacement. The blue graph shows the potential energy: it is zero in the centre of the motion and maximum when the displacement is maximum. The energy stored in a spring is proportional to the square of the extension: energy $= \frac{1}{2}kx^2$, where k is the spring constant and x is the displacement. For this reason, the energy–displacement graph for an oscillator has the (parabolic) shape of a $y \propto x^2$ graph.

The purple graph in Figure 8.3 is the total energy – the sum of the kinetic energy and the potential energy. The sum of kinetic and potential energy is constant, so the graph is a horizontal straight line.

The red graph in Figure 8.3 shows how kinetic energy varies with displacement. The kinetic energy of an oscillator is zero at the extremes of the motion, when the velocity is zero, and a maximum in the centre of the motion.

Figure 8.4 shows how energy varies for a mass executing small vertical oscillations on the end of a spring. The gravitational potential energy obviously increases from bottom to top. The energy stored in the spring decreases from bottom to top, provided the oscillations are small and the spring is always slightly stretched even at the top. Again, the kinetic energy is maximum in the middle and zero at top and bottom.

Oscillating molecules

The internal energy of a solid depends on the energy of its vibrating molecules. The molecules sometimes have kinetic energy and sometimes have potential energy. On average, half their energy is kinetic and half is potential. So the internal energy of a solid has both kinetic and potential energy components. If you increase the temperature of a solid, you increase both the kinetic energy and the potential energy of the molecules. In a liquid, too, the internal energy is both kinetic and potential.

The internal energy of gases

You read in *Electricity and Thermal Physics* that temperature is defined as that quantity which is proportional to the internal energy of an ideal gas. Atomic potential energy is due to forces between the atoms. In an ideal monatomic gas, there are no forces between the atoms, so there is no atomic or molecular potential energy. So the internal energy of an ideal gas is only kinetic.

In real gases, there are forces between the atoms and molecules. So the internal energy of a real gas is both potential and kinetic.

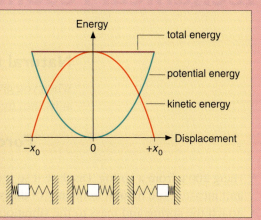

Figure 8.3 Kinetic energy is a maximum in the centre, where potential energy is zero

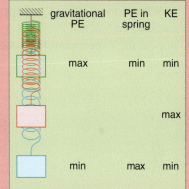

Figure 8.4 When a mass oscillates vertically on a spring, some of its potential energy is in the spring, and some is due to gravity

Mechanical resonance

Natural frequency

When you give a small displacement to a system that can oscillate, it oscillates at its own frequency. This is the oscillator's **natural frequency**.

Forced oscillations

- Hang 200 g from a spring. Displace it so it oscillates and measure the natural frequency of this system.
- Attach the spring and mass to the vibration generator as shown in Figure 9.1. Set the signal generator to produce sinusoidal waves and use a frequency meter to measure its output frequency.
- Observe the mass's motion as you slowly increase the output frequency from well below the natural frequency of the mass–spring system to well above it. What do you expect to happen to the oscillations?
- Repeat the experiment for a different oscillating mass.

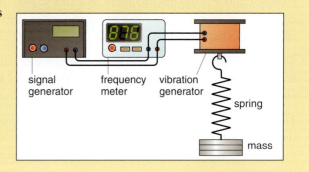

Figure 9.1   The vibrator causes the mass to oscillate

Resonance

The vibration generator is driving the mass–spring system and forcing it to oscillate. Forced oscillations are taking place. At all times, the driven system (the oscillating mass) oscillates at the frequency of the driver (the vibration generator). When the driving frequency is equal to the natural frequency of the driven system, large-amplitude, even violent, oscillations may result. This effect is called **resonance**. Resonance occurs when the driving frequency is equal to the natural frequency of the system you are driving.

Resonance curves

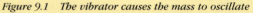

- Using the apparatus of Figure 9.1, record the maximum amplitude of the oscillating mass for a range of driver frequencies either side of resonance.
- Plot a graph of maximum amplitude against driver frequency.
- Repeat the experiment with the oscillating mass immersed in a beaker of water to provide some resistance to its motion. Repeat the measurements of maximum amplitude of oscillation for the same range of driver frequencies.
- Draw a second resonance curve on the same axes.
- How do the two graphs compare?

The effects of damping

Fluid friction, like air resistance, provides forces that reduce the amplitude of oscillations. This effect is called **damping**. For low damping, the resonance curve is sharp, and peaks when the driving frequency equals the system's natural frequency, as Figure 9.2 shows. The main effects of increased damping are to reduce the maximum amplitude of the driven system and to make resonance less noticeable. Increasing damping reduces the sharpness of the resonance curve – the peak becomes lower and wider.

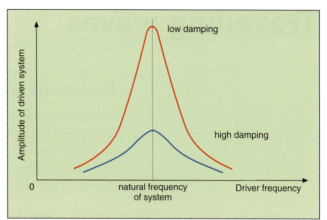

Figure 9.2   Resonance curves

Examples of mechanical resonance

There are many everyday situations that involve mechanical resonance. Some of these are useful, whereas others can be destructive and need to be avoided. In wind instruments, for example, sound is produced as a result of resonance (Figure 9.3). The vibrations of a small reed forces the molecules in an air column to vibrate.

Have you ever wondered why you find walking at one speed less tiring than walking at any other? Each of your legs is effectively a pendulum. The mass is not concentrated at a single point, so you cannot use a simple equation to find the period. But you can measure the natural frequency of your legs by standing on one leg and allowing your other leg to swing freely. Measure its natural frequency of oscillation and try to move your legs at this rate next time you go for a walk.

Figure 9.3   Resonance effects can be pleasing

Destructive resonance occurred in some early suspension bridges, which were subject to catastrophic failures when winds blowing past the structure caused oscillations at the natural frequency. Figure 9.4 shows the result of failing to prevent the frequency of these wind-induced oscillations matching the natural frequency of the bridge. Nowadays, engineers test models of such structures in wind tunnels before they are constructed, and try to modify the bridge to reduce the effects of resonance.

The parts of any machine containing a motor or an engine are subjected to periodic driving forces. Their design and construction must take account of this. Engineers try to ensure that the natural frequencies of the various parts of a car or an aeroplane are different from any periodic driving forces that the vehicle may experience during motion.

Figure 9.4   Resonance effects can be disastrous

Travelling waves

Pulses and waves

If you throw a stone into a pond, ripples spread out from where the stone hit. Disturbances like this die out; they are called wave pulses, because they last for a short time. If you want to produce continuous waves, you need to make repeated disturbances. The simplest sort of waves come from disturbances that are sinusoidal – from simple harmonic motion.

Observing waves

- Set up a ripple tank with a single dipper in the water (Figure 10.1). Tap the dipper and observe the shape of the ripples on the screen. Then use the vibrator to make continuous waves. Observe how the amplitude of the waves changes as the waves get further from their source.
- Connect a large loudspeaker to a signal generator producing a low-frequency signal. Listen to the sound. Place a lighted candle close to the centre of the loudspeaker and observe the movement of its flame (Figure 10.2). Move the candle further away and repeat the experiment. What do you observe?
- In a darkened room, connect a power supply to a small lamp. Hold a piece of paper near to the lamp and observe how brightly it is illuminated. Predict the illumination if you increase the distance between the paper and the lamp. Then test your prediction.

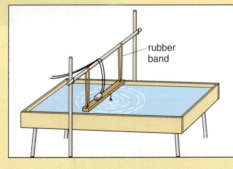

Figure 10.1   The moving dipper produces waves

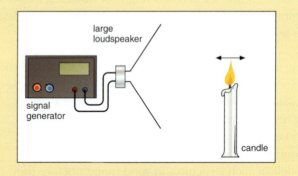

Figure 10.2   Observe the flickering flame

Energy and information

It is easy to observe the disturbances involved in water waves and sound waves. But the light emitted by the lamp is also a wave – an **electromagnetic wave** with electrical and magnetic disturbances. All these waves travel out from the source that makes them. For this reason, they are called **travelling waves** or **progressive waves**. Travelling waves from the Sun convey energy to the Earth. But they also give us information about the Sun – we can tell how hot it is, and some things about the elements that it consists of. Sound waves convey information to the listener, but also energy to the ear. This is true of all travelling waves; they convey energy and information along the direction of travel.

Variation of light intensity with distance from a point source

- You can consider the small lamp in Figure 10.3 to be a point source, radiating light uniformly in all directions.
- In a darkened room use a light meter to record measurements of light intensity at a range of distances from the lamp.
- Try to suggest a relationship between intensity and distance.

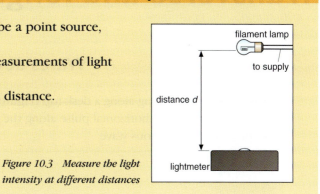

Figure 10.3   Measure the light intensity at different distances

The inverse square law

You can observe from the above experiment that waves get weaker as they spread out from a source. At a greater distance from a source, the same power is spread over a larger area, so the power per unit area is less. The power per unit area is called the **energy flux** or **intensity**, and is defined as the energy that the wave carries perpendicularly through unit area each second. The unit of energy flux is watts per metres squared (W m^{-2}). The symbol for energy flux is ϕ (the Greek letter *phi*, pronounced "fy").

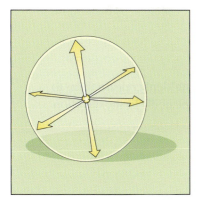

Figure 10.4   At a distance r, the light spreads over area $4\pi r^2$

If waves spread out uniformly in all directions from a **point source**, there is a simple relationship between the flux and the distance from the source. Figure 10.4 shows a point source of waves. The power of the source (the energy it emits per second) is P. At a distance r from the source, this power is spread over a sphere of radius r and area $4\pi r^2$. So the energy flux ϕ is the power per unit area, in this case given by

$$\text{energy flux} = \frac{\text{power}}{\text{area}} \quad \text{or} \quad \phi = \frac{P}{4\pi r^2}$$

This relationship is called the **inverse square law**. The energy flux will be reduced by a factor of 4 $(= \frac{1}{2^2})$ when the distance is doubled, and reduced by a factor of 9 $(= \frac{1}{3^2})$ when the distance is trebled.

WORKED EXAMPLE

A mains lamp has a light output of 12 W. What is the energy flux (intensity) 2 m from the lamp?
Assume that the lamp is a point source with its energy at any particular distance distributed over the surface of a sphere. For this example, the radius of the sphere is 2 m. The surface area of this sphere is:
surface area $= 4\pi r^2 = 4 \times \pi \times (2\text{ m})^2 = 50.3\text{ m}^2$
and the energy flux is found from

$$\text{energy flux} = \frac{\text{power}}{\text{surface area}} = \frac{12\text{ W}}{50.3\text{ m}^2} = 0.24\text{ W m}^{-2}$$

Now find the rate at which light energy would be incident on this book if it were held open 2 m from and facing the lamp. What would be the effect of increasing this distance to 4 m?

Transverse and longitudinal waves

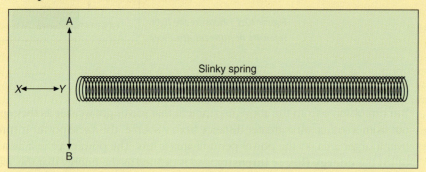

Waves down a spring

- Stretch a 'Slinky' spring along a desk top (Figure 11.1). Flick one end briefly along the line AB and back to send a single horizontal pulse along the spring. Move the end repeatedly along the same line to produce a continuous wave.

Figure 11.1   You can produce two types of wave on the spring

- Then gather together several coils at one end of the stretched 'Slinky' and suddenly release them to observe a different type of pulse moving along the spring. Move the end repeatedly in the direction *XY* to produce a continuous wave of the same type.
- Describe the difference between the two types of wave.

Transverse and longitudinal waves

You can produce two different types of waves down a 'Slinky' spring. In one type of wave, shown in Figure 11.2, the disturbances of the spring are perpendicular to the direction of travel of the wave. This type of wave is called a **transverse wave**.

If the disturbances of the spring are along the direction of oscillation, the wave is a **longitudinal wave**, as Figure 11.3 shows.

When you observe a candle in front of a loudspeaker, it shows that the oscillations associated with sound waves are parallel to the direction of travel. Sound waves are longitudinal.

It is easy to see the transverse components of water waves, because the surface of the water moves up and down as the wave spreads out. But the water also moves back and forwards as well; water waves are both longitudinal and transverse.

Electromagnetic waves such as radio waves, microwaves and light consist of electric and magnetic fields, which oscillate at right angles to each other and to their direction of travel. So electromagnetic waves are transverse.

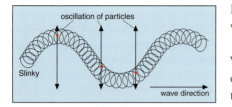

Figure 11.2   A transverse wave

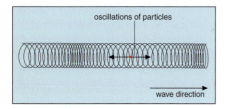

Figure 11.3   A longitudinal wave

Microwaves leaving a source

- Set up a microwave transmitter facing a receiver (Figure 11.4). Observe the strength of the received signal as you rotate the receiver through 360°.
- Repeat the experiment but leave the receiver stationary while rotating the transmitter.
- Align the receiver and transmitter for maximum signal. Place a grille of metal rods between them. What changes occur in the received intensity as the rods are rotated about the same horizontal axis?

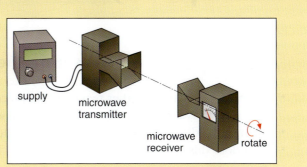

Figure 11.4   What happens when you rotate the receiver?

Plane polarisation

Microwaves are transverse waves. Those leaving the transmitter consist of oscillations restricted to one direction, determined by the structure of the transmitter. Oscillations like this, where the direction of the oscillations and the direction of travel all lie in a single plane (Figure 11.5), are called **plane polarised**. Microwave receivers are polarised; they will only receive oscillations in one direction. If the polarisation of the receiver is the same as that of the transmitter, the signal received is strong. If the polarisation of the receiver is at right angles to that of the transmitter, no signal is received. A grille of metal rods acts as a filter, letting through signals of one polarisation, but reflecting back the other components.

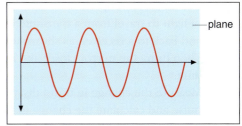

Figure 11.5   The displacements and the direction of travel define a plane

Polarising light

- Look at a lamp through a polaroid filter.
- Then look through two polaroid filters. Rotate one filter and observe what happens (Figure 11.6).
- Try to explain your observations.

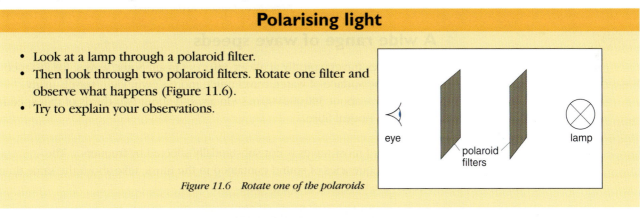

Figure 11.6   Rotate one of the polaroids

The light leaving a filament lamp is **unpolarised**; the oscillations are not confined to any single plane.

A polaroid filter polarises the light, as it confines the oscillations to a single plane, as shown in Figure 11.7. A second polaroid filter will stop the light if it is at right angles to the first filter. If it is at a different angle it will only let through a component of the polarised light from the first filter.

Both microwaves and visible light are transverse waves, and can therefore be polarised. In longitudinal waves, the oscillations are parallel to the direction of wave travel. It makes no sense, therefore, to think of confining them to a single plane. This explains why longitudinal waves, like sound, cannot be polarised.

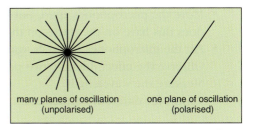

Figure 11.7   End-on view of two beams of light

Wave speed, wavelength and frequency

Measuring the speed of sound

- Watch a friend bang together two large flat pieces of wood some distance away across a sports field. Time the interval between seeing the impact and hearing the sound.
- Measure the distance that the sound has travelled and calculate its speed.
- What does this calculation assume?

Measuring the speed of light

- Use an oscilloscope to time how long it takes a flash of light to travel from a transmitter, to a distant mirror and back to a receiver, as shown in Figure 12.1.
- Then bring the mirror closer and measure the new time.
- Measure the change in travel distance of the light pulse and calculate the change in time. Use these to find the speed of light.
- Send the pulse through a length of fibre-optic cable and calculate the speed of light through that.

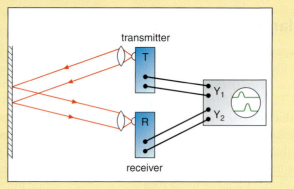

Figure 12.1   The oscilloscope times light pulses travelling from the transmitter to the receiver

A wide range of wave speeds

Sound waves travel typically at about 330 m s^{-1} in air. Some waves, for instance those on the surface of water, travel much more slowly. Light travels many times faster – about 300 000 000 m s^{-1} in air, and about two-thirds of that speed through an optical fibre.

The speed of most waves is not substantially affected by frequency. The different frequencies of sound contained in the music take about the same time to reach the back of a concert hall.

Investigating sound waves

- Set up the apparatus in Figure 12.2. Set the signal generator to give a sinusoidal output of 3 kHz. Check that the traces on the oscilloscope screen are similar when the microphones are side by side.
- Move the microphones apart, but keep them the same distance from the loudspeaker. What effect does this have on the traces on the oscilloscope's screen?
- Put the microphones together again and then move one of them further from the loudspeaker. Observe the effect. Identify two points in line with the loudspeaker that are one wavelength apart. Measure the wavelength.
- Repeat for different frequencies.

Figure 12.2   Compare the signals from the two microphones

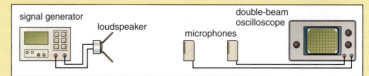

Wavefronts

Imagine freezing a wave spreading out from a single source on a water surface. Figure 12.3 shows a slice of the snapshot. Along the dotted line P – P, a fixed distance from the source of waves, all the displacements of the water are in phase. Line P – P is a **wavefront** – a line joining points of the wave that are all in phase. The line Q – Q is another wavefront, some distance further from the source. As well as being in phase with each other, the points along the line Q – Q are also in phase with those along P – P. The wavefronts P – P and Q – Q are one wavelength apart. The minimum distance between two points on a wave that oscillate in phase is called the **wavelength**, λ. It is the distance from a point on one wavefront to the corresponding point on the next wavefront, as shown in Figure 12.4.

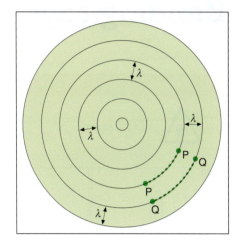

Figure 12.3   Wavefronts spread out from the source

There is a simple connection between **wave speed**, frequency and wavelength. Since frequency = number of cycles each second, and wavelength = length of each cycle, we have that

frequency × wavelength = total length each second = speed of wave

$$f\lambda = c$$

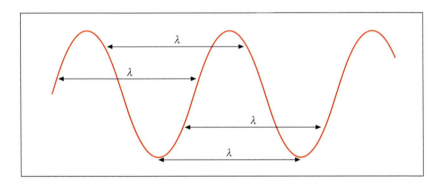

Figure 12.4   Wavelength λ

WORKED EXAMPLE

Radio 5 broadcasts on the medium wave with a wavelength of 330 m. Taking the speed of radio waves as $3.0 \times 10^8 \, \text{m s}^{-1}$, calculate the frequency and the number of oscillations that the transmitter emits during a 1 hour programme.

Since $f\lambda = c$, we get

$$f = \frac{c}{\lambda} = (3 \times 10^8 \, \text{m s}^{-1})/(330 \, \text{m}) = 9.09 \times 10^5 \, \text{Hz} = 909 \, \text{kHz}$$

During one hour there are 3600 seconds. So in 1 hour there are

$$(9.09 \times 10^5 \, \text{oscillations/second}) \times 3600 \, \text{seconds} = 3.27 \times 10^9 \, \text{oscillations}$$

Bending rays

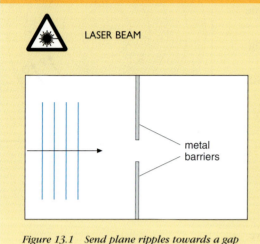

LASER BEAM

Figure 13.1 *Send plane ripples towards a gap*

metal barriers

- A ray box uses a lamp and a slit to make a narrow beam, or ray, of light. Figure 13.1 shows two metal barriers in a ripple tank. Try to use the apparatus to make a narrow ray of water ripples. Investigate the effect of changing the width of the gap in the barrier and changing the wavelength of the waves.
- In a darkened room, try to make a laser beam even narrower by passing it through a slit. Observe what happens when the slit gets very narrow.
- At a distance of 1m from a microwave transmitter, measure the width of the beam. Then put two metal plates in front of the transmitter and measure the width again. Investigate the connection between gap width and beam width.

Diffraction

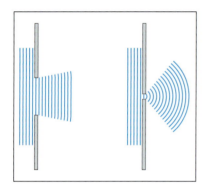

Figure 13.2 *There is greater diffraction with a smaller gap*

Whenever waves pass through a small opening, they spread out. This is called **diffraction**. The amount of diffraction depends on the size of the aperture and on the wavelength of the waves. As Figure 13.2 shows, the narrower the opening, the greater the diffraction. Also, as Figure 13.3 shows, the larger the wavelength, the greater the diffraction. Diffraction is only noticeable when the size of the opening is comparable to the wavelength. With water waves, diffraction is noticeable with openings of a few centimetres. The wavelength of light is much smaller, and diffraction of light is only noticeable with very tiny openings.

Rays and wavefronts

If you had a large ripple tank, you could make a series of flat wavefronts that would be a water **ray**. The wavefronts themselves are perpendicular to their own direction of travel; they are perpendicular to the rays.

A ray of light is also a series of wavefronts as shown in Figure 13.4. All the light wavefronts are perpendicular to the light ray. Both rays and wavefronts are ways of considering the behaviour of waves.

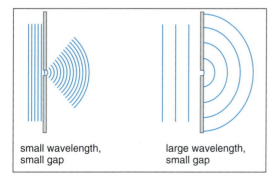

small wavelength, small gap

large wavelength, small gap

Figure 13.3 *There is greater diffraction with larger wavelength*

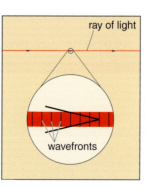

ray of light

wavefronts

Figure 13.4 *A ray is made of many wavefronts*

Reflection

When waves travel through a uniform medium, their direction of travel remains the same. But if there is a change in the medium, or some form of obstruction, the waves may be reflected. **Reflection** literally means "sent back'; waves that are reflected return to some extent in the direction from which they came. All waves can be reflected.

Reflection of microwaves

- Set up the apparatus in Figure 13.5. The metal plate between the transmitter and the receiver is used to prevent any microwaves travelling directly between them. Mark a point mid-way along the reflector and aim the microwave transmitter at it. Similarly, always aim the receiver at this point as you move it along the arc indicated.
- Adjust the position of the receiver until it detects the maximum signal. The angles of incidence and reflection should then be measured and compared.
- Repeat for a range of incident angles.

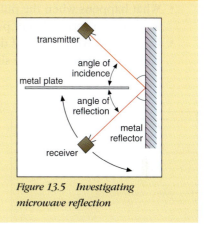

Figure 13.5 Investigating microwave reflection

Refraction

Waves in deep and shallow water

- Use a rectangular sheet of glass to create a shallow area a few millimetres deep in part of a ripple tank. Send plane waves perpendicularly towards the edge of the plate. What happens to the speed, wavelength and frequency of the ripples as they go from deep to shallow water?
- Measure the wave speed in both deep and shallow water.
- Repeat the above experiment but turn the glass plate so that the water waves approach it at an angle, as shown in Figure 13.6. Sketch the pattern of the wavefronts either side of the boundary between the deep and shallow water.

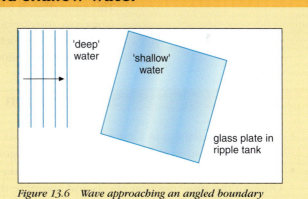

Figure 13.6 Wave approaching an angled boundary

When a wave crosses into shallower water, it slows down. The wavefronts in the deep water behind are travelling faster. They catch up with slower waves in front and get closer. So both the wavelength and the speed are reduced in the shallow water.

In the shallow water, the waves are slower and closer together. In the deep water, the waves are faster but further apart. In both the deep and shallow water, the number of waves that pass a point in each second – the frequency – remains the same.

When a wave crosses into shallower water at an angle, the wavefront changes its direction of travel towards the normal. This change in direction, resulting from a change in speed at the boundary, is called **refraction**.

The principle of superposition

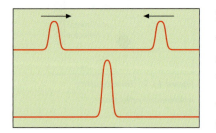

Figure 14.1 The pulses add up where they cross

Superposition

When two pulses travelling in opposite directions on a spring meet, they pass straight through each other. As Figure 14.1 shows, while they are crossing, the combined pulse is the addition of the two separate pulses.

The displacements of the pulses are vector quantities. Their resultant takes into account their direction. If the pulses have the same direction, they add up to give a larger resultant displacement. If the pulses have opposite directions, they totally or partially cancel out.

This same effect is evident with continuous waves. When two waves of the same type cross, the displacement at any point is equal to the vector sum of the displacements of each of the waves at that point. This is called the **principle of superposition**. (Superposition means 'placing on top'.)

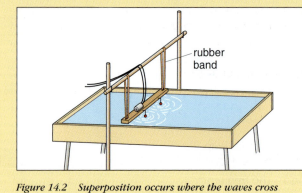

rubber band

Figure 14.2   Superposition occurs where the waves cross

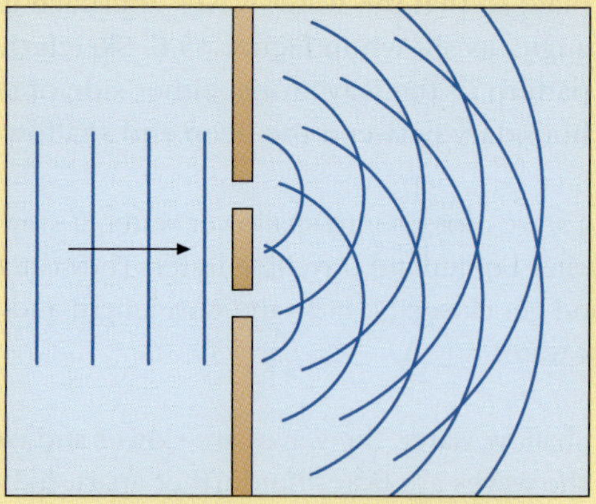

Figure 14.3   You can use two gaps to make two sets of waves

Superposition patterns

You can produce sets of circular waves either with separate dippers or using diffraction through holes in a barrier. However you do it, the two overlapping sets of circular water waves produce a complicated pattern like that in Figure 14.4. In some places the directions of the varying wave displacements from the two sources are the same. The waves are in phase, so they reinforce each other. This is **constructive superposition** (sometimes called constructive interference). In other places the directions of the varying wave displacements from the two sources are opposite. The waves are out of phase, so they cancel. This is **destructive superposition** (or destructive interference).

Where the waves arriving at a point are in phase, there is constructive superposition. When the waves are in antiphase (π out of phase), there is destructive superposition.

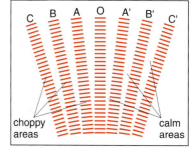

Figure 14.4   A superposition pattern on water shows calm areas and choppy ones

Using separate sources

- Use the two separate vibrator and dipper assemblies to make two sets of circular waves (Figure 14.5). Try to set both dippers vibrating at the same frequency. Observe the resulting superposition pattern.
- With the vibrating frequencies the same, start and stop one vibrator. Observe what happens to the pattern.
- What happens to the pattern if the frequencies are different?

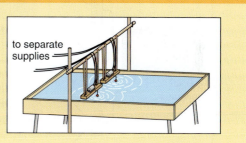

Figure 14.5   These dippers vibrate independently

Coherence

Two sources of different frequencies will not produce a stable superposition pattern. Places where the waves are in phase at one instant will become out of phase the next. If the two sources have very close frequencies, you may be able to see a pattern that moves slowly. If the frequencies are very different, the pattern will move so quickly that it will be impossible to see.

For superposition patterns to be stable and observable, the waves that produce them must be of the same frequency.

If the sources producing a superposition pattern have the same frequency, but are out of phase with each other, a stable pattern will be produced. But if one or both keeps starting and stopping, so that they are sometimes in phase with each other and sometimes out of phase, then the pattern will not be stable.

For a stable superposition pattern, the sources of waves must be **coherent**. This means that they must have the same frequency, and any phase difference must be constant.

The effect of amplitude on superposition

If two sources of waves have very different amplitudes, then you will not observe superposition patterns. The wave with the larger amplitude will dominate, because it makes little difference to the total whether the wave with the smaller amplitude is in phase or out of phase with it. When the waves have similar amplitude, cancellation is noticeable. And if the amplitudes are identical, then cancellation can be complete.

Two-source superposition experiments

Superposition of sound waves

- Set up the apparatus of Figure 15.1 on a large sheet of paper. The two loudspeakers connected to the single signal generator produce two coherent sets of sound waves.
- Adjust the output of the signal generator to 3 kHz. Locate positions of maximum and minimum sound intensity.
- Check that there is constructive superposition along a line equidistant from both speakers.
- Locate and mark the other lines of constructive superposition and the lines of destructive superposition.

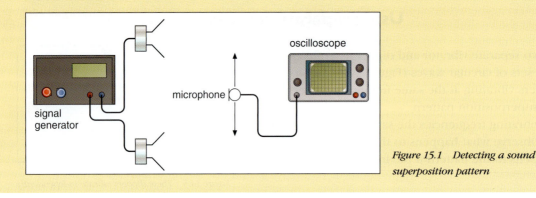

Figure 15.1 Detecting a sound superposition pattern

Path difference and phase difference

Figure 15.2 shows the superposition pattern produced by two loudspeakers vibrating in phase. It is the same as the pattern produced by two sources of water waves.

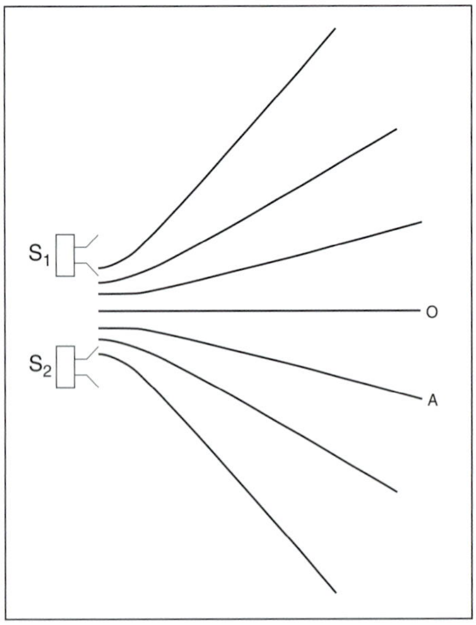

Along the central line O there is constructive superposition. Waves from S_1 and S_2 travel the same distance (the same path length) to any point on the line. So the waves are in phase along that line. There is no phase difference and hence there is constructive superposition. This central line is called the **central maximum**.

At any point along the curved line A, the path length of the wave from S_1 is a whole wavelength longer than the path length of the wave from S_2. The **path difference** is a whole wavelength; the waves arriving on the line are 2π out of phase, which is like being in phase, so there is still constructive superposition. Midway between lines O and A there is a region of destructive superposition where the path difference is half a wavelength and the phase difference is π. The waves are out of phase and they cancel.

Figure 15.2 The lines show the regions of constructive superposition

Superposition with microwaves

- Set up the apparatus in Figure 15.3. Diffraction at the two gaps in the metal plates produces two coherent sets of microwaves from the single transmitter. The amplifier and loudspeaker indicate the strength of the microwaves received by the probe.
- Mark the position of each slit on the paper. Move the detector along the line shown. Mark and label the positions of maximum and of minimum signal strength on the sheet of paper.
- Measure the distance from the centre of each slit to the points of maximum and minimum signal strength. Use these distances to find an average value for the wavelength of the microwaves.

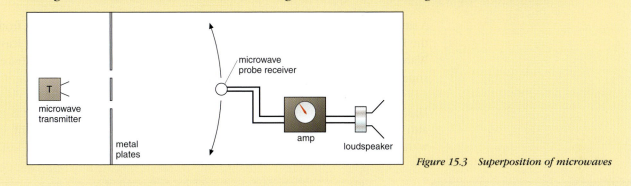

Figure 15.3   Superposition of microwaves

Path difference

Figure 15.4 shows two sources of microwaves at S_1 and S_2. The microwaves leave the two slits S_1 and S_2 in phase. The point O is the same distance from both slits. The path difference is zero. The waves are in phase when they reach O and so there is constructive superposition. O is on the central maximum.

The distance from S_2 to A is greater than the distance from S_1 to A by an amount λ. So the waves arriving at point A are in phase. There is constructive superposition. Midway between O and A, the path difference is $\lambda/2$. The waves are out of phase, so there is destructive superposition. At point B, the path difference is 2λ; there is constructive superposition. Midway between A and B, the path difference is $1\frac{1}{2}\lambda$, and there is destructive superposition.

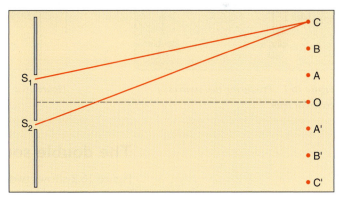

Figure 15.4   A maximum occurs at C when $S_2C - S_1C = 3\lambda$

In general:

For constructive superposition, path difference $= n\lambda$.

For destructive superposition, path difference $= (n + \frac{1}{2})\lambda$.

The series of maxima and minima are known as *fringes*.

Superposition of light

Producing coherent light sources

Light sources, even monochromatic ones, produce bursts of waves with different phase relationships, rather than a continuous coherent wave. The light from a filament lamp is particularly incoherent, as each part of the filament produces its own set of incoherent waves.

A small slit in front of a filament lamp samples only a fraction of the light from the filament and diffracts it. That diffracted light is coherent with itself at least. With a filter to make it monochromatic, the light can be used to demonstrate superposition, as shown in Figure 16.1.

Light from a laser is already monochromatic and much more coherent. It is also much brighter. It is much easier to demonstrate superposition of light with a laser.

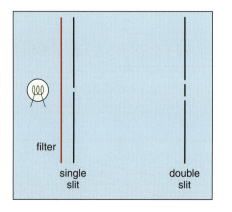

Figure 16.1   Producing two sources of coherent light

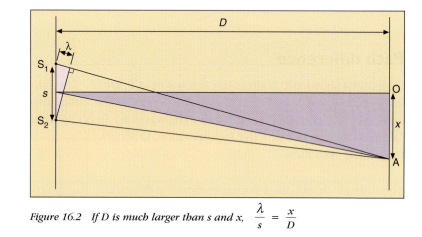

Figure 16.2   If D is much larger than s and x, $\dfrac{\lambda}{s} = \dfrac{x}{D}$

The double source equation

Figure 16.2 shows two wave sources, S_1 and S_2, a distance s apart. They produce a central maximum at O, and the first maximum next to that (the first **subsidiary maximum**) a distance x away at point A. If distance D is much larger than s and x, then the two shaded triangles are similar. So

$$\frac{\lambda}{s} = \frac{x}{D} \qquad \text{so} \qquad \lambda = \frac{xs}{D}$$

You can use this formula to measure the wavelength of light. The distance x is the fringe width. Not only is it the spacing between the central maximum and the first maximum at one side, it is also the spacing between any two adjacent fringes near the centre of the superposition pattern.

Young's double slit experiment

- Use a laser to illuminate two narrow slits about 0.5 mm apart. Light diffracts at the slits and the two overlapping sets of waves produce a superposition pattern on the screen (Figure 16.3).
- Measure the width of several fringes and calculate x, the fringe width.
- Measure also the slit separation and the distance between the screen and the slits. Use these, and the formula $\lambda = \dfrac{xs}{D}$, to calculate the wavelength of the laser light.

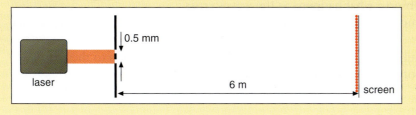

LASER BEAM

Figure 16.3 Typical arrangement when using a laser

Monochromatic and chromatic fringes

Figure 16.4 shows the superposition pattern produced by red laser light. The slit spacing was 0.14 mm and the slit-screen distance 3.0 m. You can use the scale marked on the photograph to calculate the wavelength of the light. Each division represents 1 cm.

The wavelength of violet light is about 4×10^{-7} m (400 nm) whereas that of red light is about 7×10^{-7} m (700 nm). So red light gives a larger fringe spacing than blue light when both are used in the same double slit experiment. If slits are illuminated by both red and blue light simultaneously, the red fringes start getting mixed up with the blue fringes.

Figure 16.5 shows the fringe pattern produced by white light. You can see that the first subsidiary blue maximum is close to the central maximum, and the first subsidiary red maximum is further from the central maximum. This produces a blue tinge to one edge of the first subsidiary maximum and a red tinge to the other edge. Notice how the fringes become less distinct the further out they are from the central maximum, as the differently coloured fringes overlap more.

Figure 16.4 Superposition pattern for laser light through a double slit

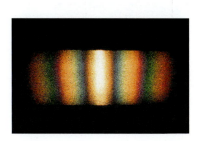

Figure 16.5 Superposition pattern caused by white light through a double slit

WORKED EXAMPLE

A yellow lamp produces a wavelength of 550 nm. Light from this lamp passes through a single slit and illuminates a double slit, which has a slit separation of 0.45 mm. A screen is placed 2.1 m away from the double slit. Calculate the fringe width.
Since

$$\text{wavelength} = \frac{\text{fringe spacing} \times \text{slit separation}}{\text{distance to screen}}$$

we have

$$\text{fringe spacing} = \frac{\text{wavelength} \times \text{distance to screen}}{\text{slit separation}}$$

so

$$\text{fringe spacing} = \frac{550 \times 10^{-9}\,\text{m} \times 2.1\,\text{m}}{0.45 \times 10^{-3}\,\text{m}}$$

$$= 2.6 \times 10^{-3}\,\text{m} = 2.6\,\text{mm}$$

Stationary waves

- Stretch a long 'Slinky' spring between yourself and a friend. Simultaneously send waves of the same frequency from both ends. Observe what happens as the waves cross.
- Adjust the frequency to produce a regular pattern of oscillation. Sketch this pattern. Try to produce other patterns.
- Fix one end of the spring firmly and send waves from the free end. What happens to these waves when they reach the far end of the spring? Again, try to produce a range of patterns.

Waves in opposite directions

There are two ways of investigating the effect of identical waves travelling in opposite directions on a spring. Either you can produce both waves directly, sending them from opposite directions down a spring, or you can send one wave down and observe it crossing its own returning reflection.

When identical travelling waves cross in this way, they produce a wave shape that stands still on the spring, called a **stationary wave**.

Nodes and antinodes

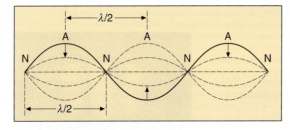

Figure 17.1   A stationary wave

Figure 17.1 shows a stationary wave on a spring. Some points on the spring, labelled N, always remain stationary. At these places, called **nodes**, the amplitude of oscillation is zero. At other places, labelled A, the spring vibrates with maximum amplitude. These places are called **antinodes**. Adjacent nodes, or adjacent antinodes, are half a wavelength apart.

- Direct a microwave transmitter at an aluminium plate reflector (Figure 17.2). Move a probe receiver along the line between the transmitter and the plate, and observe the nodes and antinodes.
- Measure the average distance between several adjacent nodes, and use this to calculate the wavelength of the microwaves.

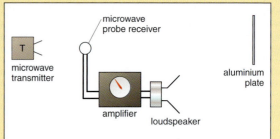

Figure 17.2   The forward and reflected microwaves produce a stationary wave

Constructive and destructive superposition

Nodes are places of no displacement. Nodes occur when the varying displacements of the two waves arriving at a point from opposite directions always cancel out. The wave displacements are equal and opposite to each other at all times. When the wave that is moving to the right through a node is at its positive maximum, the wave moving to the left is at its negative maximum. Destructive superposition occurs. Any displacement of one of the waves is cancelled out by an equal and opposite displacement of the other wave.

Antinodes are places of maximum amplitude. They occur when the two waves arriving at a point are in phase. Constructive superposition takes place, and the amplitude is twice that of each individual wave.

Energy and stationary waves

A travelling wave conveys energy in the direction of travel. A stationary wave consists of two identical travelling waves going in opposite directions. You might imagine that this conveys equal energies in opposite directions. The net effect of this is that there is no energy transfer. A stationary wave conveys no energy at all.

Waves on a tensioned rubber cord

- Stretch a rubber cord between two fixed points. Use a signal generator to set the cord vibrating at a frequency of 1 Hz (Figure 17.3). Gradually increase the frequency until there is a stationary wave pattern. Sketch this pattern. Measure the wavelength and record its frequency.
- Continue to increase the frequency slowly until the next stationary wave pattern is produced. Again, sketch this and record its frequency and wavelength.
- Repeat this procedure for the first five stationary wave patterns. How do the frequencies of the different patterns compare?
- Use your results to find the speed of waves on the rubber cord.

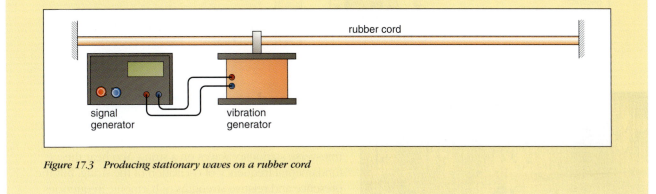

Figure 17.3   Producing stationary waves on a rubber cord

Oscillations of a stationary wave

The first, or fundamental, mode of oscillation of the stationary wave on the rubber cord occurs at the **fundamental frequency** f. As Figure 17.4 shows, at this frequency the wavelength is twice the length of the cord. The cord has a node at each end, with a single antinode in the middle.

STATIONARY WAVES

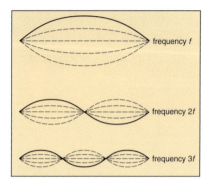

Figure 17.4   Different modes of vibration

The next stationary wave pattern is observed at *2f*, twice the fundamental frequency. The wavelength is half as much, and twice as many node-to-node loops can fit onto the cord. You can get further stationary wave patterns at integer multiples of the fundamental frequency. These are the **harmonic frequencies** (*2f*, *3f*, *4f*, etc).

A larger frequency is required to produce each pattern if the tension in the rubber cord is increased. The wavelength remains fixed if the length of the cord remains constant, indicating an increase in the wave speed with tension.

Phase along a stationary wave

All particles in a medium between two adjacent nodes move in the same direction together; they vibrate in phase with each other. Particles either side of a node move in opposite directions; they are vibrating out of phase with each other.

Stationary sound waves

- Point a loudspeaker at a hard wall. Set the signal generator to give a frequency of about 3 kHz. Use the microphone and oscilloscope to locate the nodes and antinodes (Figure 17.5). Measure the frequency and the wavelength.
- Change the frequency to obtain a different stationary wave pattern. Record a series of corresponding readings of frequency and wavelength.
- Plot a graph of frequency against $\dfrac{1}{\text{wavelength}}$. Use the gradient of this graph to find the speed of sound in air.

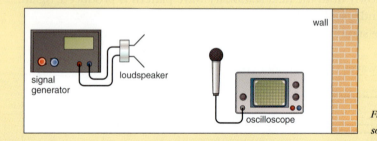

Figure 17.5   Producing stationary sound waves

Music and stationary waves

Figure 17.6   Stationary waves produce most orchestral sounds

Most musical instruments use stationary waves to produce their notes (Figure 17.6). Stringed instruments use the transverse stationary waves on a string. You hear the fundamental frequency of oscillation of the string, but the presence and relative strengths of harmonic frequencies present gives each musical instrument its own characteristic sound. Wind instruments rely on the production of longitudinal stationary waves in columns of vibrating air.

Quantum physics

An exciting place to be, particularly for atoms!

Photoelectric emission

- An electroscope has a thin gold leaf attached to a metal stem. When the electroscope is charged, the leaf stands out.
- Use an EHT power supply to charge an electroscope negatively (Figure 18.1). Disconnect the power supply and watch the electroscope for a minute. Does it discharge?
- Repeat the experiment, but this time shine ultra-violet radiation on to the electroscope. Is there a difference?
- Repeat the experiment yet again, but this time put a clean strip of zinc on the top cap (Figure 18.2).
- Experiment with a positively charged electroscope, with strong visible light, and with dull, oxidised zinc.

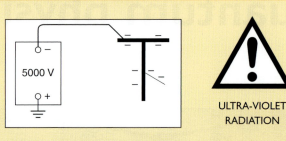

Figure 18.1   Charging an electroscope

ULTRA-VIOLET RADIATION

Figure 18.2   Put a cleaned strip of zinc on the electroscope

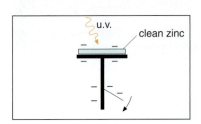

Figure 18.3   The leaf goes down

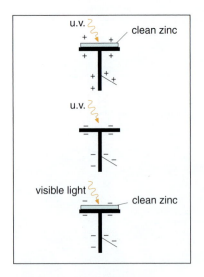

Figure 18.4   The leaf stays up

Photoelectric emission

When you connect the negative terminal of a power supply to an electroscope, some electrons transfer to the electroscope, making it negative too. The leaf on the electroscope then has the same sign of charge as the stem, so the two repel each other and the leaf stands out. The insulation on an electroscope is usually good, so very little charge leaks away and the leaf stays out. But, in certain circumstances, when you illuminate the top cap with the correct kind of radiation, an electroscope with a good insulator can still discharge rapidly.

In the situation shown in Figure 18.3, the leaf goes down, even with nothing touching the electroscope. The leaf discharges only if the zinc is shiny, if the illumination is ultra-violet and if the charge is negative. It does not discharge if the zinc is dull or if there is no zinc, with just bright light, or with positive charge (Figure 18.4).

You might at first think that the ultra-violet light is causing the air around the electroscope to conduct electricity, and so discharge the electroscope. But if so, the electroscope would also discharge if it were charged positively. If ionisation of air were the explanation, why should a zinc plate be necessary? Perhaps the ultra-violet light might be giving positive charge to the electroscope to neutralise it. But if so, why is there discharge when the zinc plate is on the top cap, but no discharge with a chromium top alone?

A likely explanation for the discharge is that ultra-violet radiation is causing the zinc to emit electrons and make itself less negative. This emission is **photoelectric emission** (literally 'emission of electrons by light').

Threshold frequency

Photoelectric emission from zinc only occurs if the radiation illuminating the zinc has a frequency of 1×10^{15} Hz or higher. This is in the ultra-violet region, just outside the visible spectrum. This frequency is called the **threshold frequency** for zinc.

Weak radiation above the threshold frequency will cause photoelectric emission, but radiation below the threshold frequency will not cause photoelectric emission, even if it is powerful. This observation puzzled physicists for some years.

Photons

In 1905, Einstein suggested an explanation for photoelectric emission based on a theory proposed by Max Planck (Figure 18.5). He suggested that electromagnetic radiation – visible light, ultra-violet light, or any other frequency – comes in small packets of energy, rather than in a steady stream. The general name for a small packet of energy is a **quantum**, but a packet, or quantum, of electromagnetic radiation is called a **photon**. The energy of a photon does not depend on the intensity of the radiation, but rather on its frequency.

When the frequency of the radiation is low, the energy of the photons is small; when the frequency is high, the energy of the photons is large.

The electrons in the metal are being bombarded with a stream of photons. An electron is only emitted if it interacts with a photon that has sufficient energy, on its own, to detach the electron from the metal. When photons of lower energy hit the metal, no electrons are emitted.

Figure 18.5. Max Planck (1858–1947). German physicist who proposed that electromagnetic radiation is emitted in quanta

Work function

There is no photoelectric emission from zinc unless it is illuminated with radiation of frequency greater than its threshold frequency of about 1×10^{15} Hz. If you investigate other substances, you find that they each have a different threshold frequency.

Generally, the threshold frequencies are lower for substances that are chemically more reactive. These substances lose electrons more easily both in chemical reactions and photoelectrically. The lower threshold frequency corresponds to photons of lower energy; it means that you do not need photons of such high energy to release electrons from more reactive substances.

The minimum energy needed to remove electrons from a substance is called the **work function**, symbol ϕ (the Greek letter *phi*). The work function for zinc is less than the work function for chromium, which is why the threshold frequency for zinc is lower than the threshold frequency for chromium. This explains why radiation that causes photoelectric emission from the zinc plate does not necessarily cause emission from the chromium cap.

Einstein's photoelectric equation

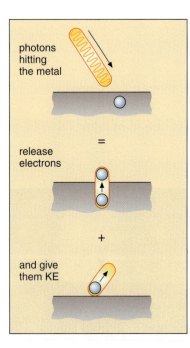

Figure 19.1   Photon energy = work function + electron kinetic energy

The energy of photoelectrons

When a photon ejects an electron, some of the photon's energy is used to free the electron from the material, and the rest gives kinetic energy to the electron (Figure 19.1). The easiest electrons to eject are those near the surface of the material. They need only the work function to release them, so there is most photon energy left over to give these electrons kinetic energy. For these, the fastest electrons, the photon energy is equal to the work function plus the electron kinetic energy.

Other electrons need more energy than the work function to eject them. When a photon ejects these electrons, they have less kinetic energy than the fastest electrons. So, with a given substance and given frequency of illumination, photoelectrons are emitted with a range of kinetic energies. But these energies have a clear maximum, which is the energy of the electrons that are easiest to remove from the metal. So if you measure the kinetic energy of the fastest emitted electrons, you can find out about both the photon energy and the work function of the metal.

To measure the energy of the electrons, you provide an electrical hill for them to run up. It is rather like measuring the energy of a bullet by finding how high it will rise when you shoot it upwards.

Measuring the energy of a photoelectron

- The photoelectric cell in Figure 19.2 has two electrodes. Light illuminates the large emitting electrode, which has a low work function. Photons with sufficient energy cause photoelectric emission. If any photoelectrons reach the receiving electrode, the picoammeter indicates a current.
- The battery and potentiometer make the receiving electrode in the photoelectric cell negative. This provides an electrical hill that photoelectrons must run up. The trick is to increase the repelling voltage slowly from zero until the current drops to zero. At this voltage, called the stopping voltage (or stopping potential), the electrical hill is just high enough to stop even the fastest electrons arriving.
- At the stopping voltage, the kinetic energy lost by the fastest electrons is equal to the electrical potential energy they gain going up the hill.
- Shine lights of different known frequencies onto the emitting electrode. For each frequency, measure the stopping voltage.
- Plot a graph of stopping voltage against frequency. What is the relation between them?

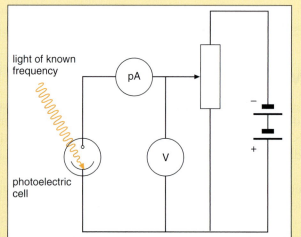

Figure 19.2   The electrons are repelled back to the emitting electrode

ULTRA-VIOLET
RADIATION

Calculating photon energy

Einstein accepted Planck's hypothesis that the energy of each photon is proportional to its frequency:

$$E \propto f \qquad \text{or} \qquad E = hf$$

The constant of proportionality h is Planck's constant; it is $6.63 \times 10^{-34}\,$J s. Since

photon energy = work function + kinetic energy of fastest electrons

$$hf = \phi + \tfrac{1}{2}mv^2$$

where ϕ is the work function, and m and v are the mass and speed of the fastest electrons.

The voltage between the electrodes is equal to the potential energy per unit charge (see *Electricity and Thermal Physics*, Chapter 7). So the potential energy gained by an electron, as it runs up the hill repelling it back to the emitting electrode, is equal to its charge multiplied by the potential difference. If the electron just fails to reach the receiving electrode, then the potential energy it gains must be equal to the kinetic energy it was emitted with:

potential energy gained = kinetic energy lost

$$QV_s = \tfrac{1}{2}mv^2$$

where Q is the charge on an electron and V_s is the **stopping voltage**. So our equation above becomes

$$hf = \phi + QV_s$$

This is Einstein's photoelectric equation. The equation can be rearranged in the form $y = mx + c$, as follows:

$$V_s = \left(\frac{h}{Q}\right)f - \frac{\phi}{Q}$$

A graph of V_s against f produces a straight line of gradient h/Q. The charge Q on an electron is known to be $1.63 \times 10^{-19}\,$C, so you can calculate Planck's constant.

Figure 19.3 Stopping voltage for caesium for a range of frequencies

Finding the threshold frequency

Look at the stopping voltage–frequency graph for caesium in Figure 19.3. As you would expect, it is a straight line. As the frequency of the radiation increases, the stopping voltage gets greater, meaning that electrons are emitted with greater energy. As the frequency gets less, the stopping voltage gets smaller, until, when the line cuts the frequency axis, electrons are emitted with zero kinetic energy. This is the **threshold frequency**. Below this frequency, no electrons are emitted.

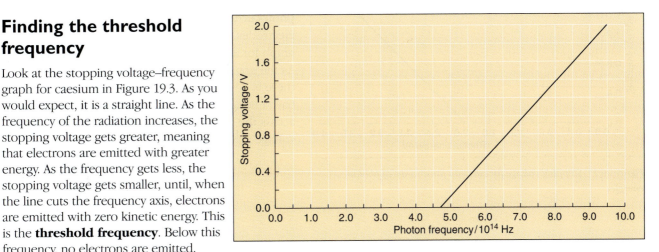

Table 20.1 *Threshold frequency for a number of metals*

Metal	Threshold frequency /10^{15} Hz
caesium	0.47
sodium	0.57
zinc	0.88
chromium	1.08
iron	1.12
copper	1.13

Table 20.2 *Work function for a number of metals*

Metal	Work function /10^{-19} J	/eV
caesium	3.11	1.94
sodium	3.78	2.36
zinc	5.81	3.63
chromium	7.10	4.44
iron	7.36	4.60
copper	7.44	4.65

The electronvolt

The work done by a charge Q moving through a potential difference V is QV. When an electron, of charge 1.6×10^{-19} C, moves through a potential difference of 1 V, the work done is given by

$$\text{work done} = QV = 1.6 \times 10^{-19}\,\text{C} \times 1\,\text{V} = 1.6 \times 10^{-19}\,\text{J}$$

This amount of work, 1.6×10^{-19} J, is called the **electronvolt**. It is the work done when a charge equal to that on an electron moves through a potential difference of 1 V.

To convert from joules to electronvolts, divide by 1.6×10^{-19} J eV^{-1}. For example, the work function of zinc, $\phi(\text{zinc})$, is

$$\phi(\text{zinc}) = 5.8 \times 10^{-19}\,\text{J} = \frac{5.8 \times 10^{-19}\,\text{J}}{1.6 \times 10^{-19}\,\text{J eV}^{-1}} = 3.6\,\text{eV}$$

The electronvolt is a conveniently sized unit of energy for photoelectricity. Tables 20.1 and 20.2 show threshold frequency and work function in both joules and electronvolts for a number of metals.

Figure 20.1 is the same graph as Figure 19.3, but the x-axis is labelled with photon energy in electronvolts instead of photon frequency. The intercept on the x-axis shows that the minimum photon energy for emission is 1.94 eV, so the work function for the emitting electrode is also 1.94 eV. For each electronvolt increase in the photon energy, the stopping voltage increases by one volt, showing that the emitted electron has an extra electronvolt of energy.

Variation of photocurrent with voltage

So far we have discussed varying the frequency of the light illuminating a photocell. If, instead, you keep the frequency constant, you can measure how photocurrent depends on the voltage between the electrodes. Figure 20.2 shows a current–voltage graph for a photocell illuminated with dim red light. As the graph shows, there is a current even when there is no voltage across the cell. The electrons have enough energy when they are emitted to travel across the gap between the electrodes even with no voltage across the cell.

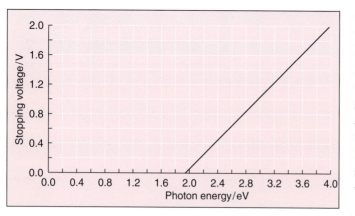

Figure 20.1 Stopping voltage for caesium for a range of photon energies

At constant illumination, electrons are emitted at a constant rate, but with varying energies. With zero voltage, some will get across to the receiving electrode. Changing the voltage helps or hinders the flow. If the receiving electrode is very positive, it gives maximum help to get all the electrons across the gap, but does not cause any more to be emitted. This is **saturation**, when all the electrons that are being emitted are being received.

If the receiving electrode is negative of the emitting electrode by an amount of the stopping voltage V_s or more, it stops even the fastest electrons moving across and the current is zero.

Figure 20.3 shows the current–voltage graph for both bright and dim red light. The stopping voltages for both radiations are the same, showing that, for both radiations, electrons are emitted with the same maximum energy. This is because both the dim and bright red light have the same frequency and therefore the same photon energy.

The line for the brighter light shows a larger saturation current. More photons are arriving per second, so more electrons are emitted per second, enabling a larger maximum current.

Figure 20.4 shows current–voltage graphs for a photocell illuminated by red and blue light of the same intensity. The stopping voltage for blue light is greater than that for red, showing that the photons of blue light have a higher energy than the photons of red.

The saturation current for blue light is less than that for red, showing that there are fewer electrons emitted per second by the blue. Both radiations have the same intensity. The blue light comes as a smaller number of photons, each of which has a larger energy, emitting fewer electrons but with higher energy. The red light comes as a larger number of photons, each of which has a smaller energy, emitting more electrons but of lower energy.

Light – wave or particle?

These three chapters on quantum phenomena have shown that electromagnetic radiation behaves as though it is a stream of photons – a stream of particles. In earlier chapters, you studied the way that light behaves as a wave. This leads to something of a puzzle: Is light a wave or a particle? Perhaps the nearest simple explanation is to say that light is made of particles that have wave properties. The wave properties of the particles describe the places in which you are likely to find the particles. When two light waves interfere destructively at a point, it is not that two sets of particles arrive at the point and cancel out, but rather that the wave sets of a single particle cancel out so there is no chance of finding a particle at that point. You can read in Chapter 23 how electrons and other particles have both wave and particle properties.

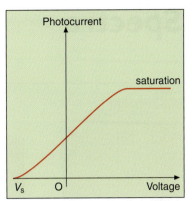

Figure 20.2   Current–voltage graph for a photocell illuminated with dim red light

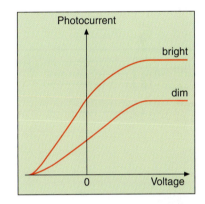

Figure 20.3   Current–voltage graphs for a photocell illuminated with two different intensities of red light

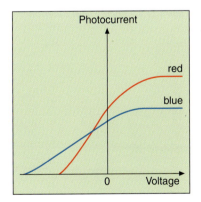

Figure 20.4   Current–voltage graphs for a photocell illuminated with different frequencies of light

Spectra

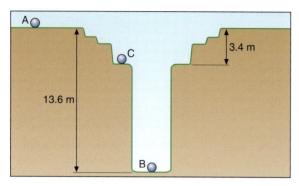

Figure 21.1   The balls have different potential energies

Stuck in a hole

Figure 21.1 shows three balls, each of weight one newton (1 N), near a hole, which has a number of levels. Ball A is free from the influence of the hole. If you use ground level as the zero for potential energy, this ball has no potential energy. Ball B is at the bottom of the hole. It is 13.6 m below ground level. It needs 1 N × 13.6 m = 13.6 J to free it from the hole. Its potential energy is 13.6 J less than that of ball A, which has an energy of 0 J; this is –13.6 J. Ball C is on one of the ledges around the hole. It is 3.4 m below the ground and has a potential energy of –3.4 J.

If ball C drops to the bottom of the hole, it loses 10.2 J, and its energy becomes –3.4 J – 10.2 J = –13.6 J.

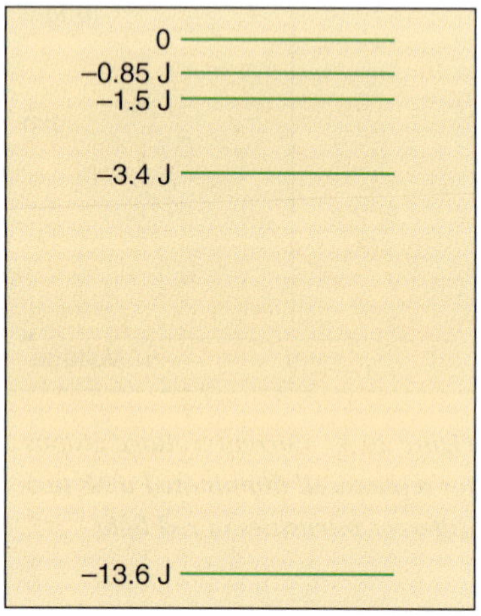

Figure 21.2 shows an energy level diagram for the hole. The diagram ignores the shape of the hole, and just shows the levels of energy that a ball in the hole can have; no other values are possible.

Figure 21.2   Energy level diagram

Electron energy levels

Electrons are held by atoms rather like the balls are held by the hole in Figure 21.1. The electron can be entirely free from the atom, in which case it is said to have zero potential energy. Or the electron can be fastened as closely as it can get to the atom. In between, there is a range of possible energy levels. In all of these energy levels, electrons have less total energy than if they were free. So these energy levels are labelled with negative values. You can use an energy level diagram to describe the possible energy states of the atom. Figure 21.3 shows an energy level diagram for the hydrogen atom, with different energies marked in electronvolts.

Normally the electron of a hydrogen atom lies in the lowest energy level – the **ground state** – like the ball lying at the bottom of the hole. But the electron can occupy any of the other possible states.

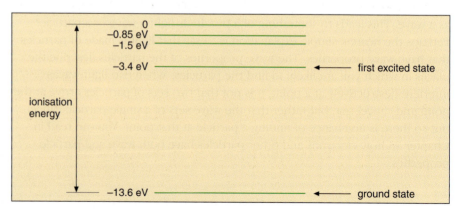

Figure 21.3   The energy levels for a hydrogen atom

Excitation and ionisation

If you give an atom energy to raise the electron above the ground state, the atom becomes **excited**. Energy is required to raise the electron above the ground state. This is **excitation energy**. The electron may remain above the ground state temporarily, but it will usually drop back to the ground state, either directly or via another level, giving out the excitation energy as it does so.

The first excitation energy for hydrogen is 10.2 eV, because it needs 10.2 eV to raise the atom from its ground state to the first excited state.

If you give the atom enough energy, you can free the electron completely from the atom. This is called **ionisation**. The **ionisation energy** is the energy required to free an electron completely, starting from the ground state of an atom. It is 13.6 eV for the hydrogen atom.

Allowable changes

The energy levels for an atom are fixed. This means that only certain transitions (changes) of energy are possible. To move between any two of these fixed levels means that the electron needs to receive, or give out, a defined amount of energy. When an atom changes from one level to a lower one, the surplus energy E is given out as a single photon of radiation. The defined changes of energy mean that there are defined frequencies of radiation given out, since $E = hf$. These frequencies therefore give information about the energy levels in an atom.

The change in energy E of an electron is the difference in energy of the two states between which the electron moves, $E = E_2 - E_1$. The frequency of the emitted radiation is given by the formula $E = hf$. So

$$hf = E_2 - E_1$$

Big jumps mean large energy changes, which correspond to high frequencies of radiation with short wavelengths. Small jumps mean small energy changes, which correspond to low frequencies of radiation with long wavelengths.

The atoms of each element have a characteristic set of energy levels, and so give out a characteristic set of frequencies of radiation when they are excited. So you can identify an element from the frequencies of the radiation it emits when excited.

Observing spectra

- Put a slit in front of a hydrogen lamp. Hold a diffraction grating next to your eye and look through it at the slit (Figure 21.4).
- Sketch the pattern of light you see.
- Repeat with lamps containing different elements.

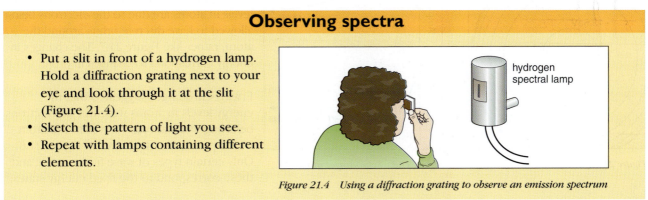

Figure 21.4 Using a diffraction grating to observe an emission spectrum

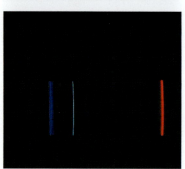

Emission spectra

You can give an element energy to excite the atoms by heating it. In this way, electrons are continually being given energy to enable them to rise to a higher state, but they then fall down again. The atoms give out the range of frequencies characteristic of that element. This range of frequencies of emitted radiations is called the **emission spectrum** of the atom.

Chemists can identify many metals just by putting a sample in a flame and identifying the characteristic spectrum by eye.

Using a diffraction grating to observe spectra

A diffraction grating works on the same principle as Young's slits (see Chapter 16). But the grating has more slits and they are closer together, so the fringes are brighter and spaced much further apart. You can use a diffraction grating to observe and measure the wavelengths of radiation from an emission spectrum. Figure 21.5 shows how you can observe the fringes that a diffraction grating can produce on the retina of the eye.

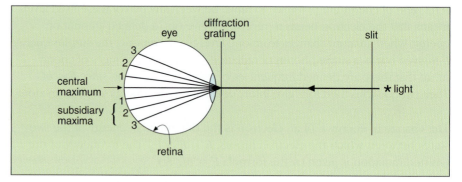

Figure 21.5   Superposition fringes on the back of the eye

Figure 21.6   Emission spectra from (top) cadmium, (centre) sodium, (bottom) hydrogen

Figure 21.6 shows the emission spectra of some common elements, including that of hydrogen. Notice the series of discrete lines, which correspond to the series of discrete energy levels in the atom.

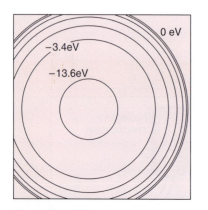

Figure 21.7   Bohr atom

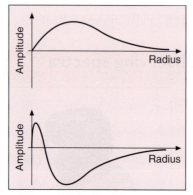

Figure 21.8   Schrödinger atom

Stationary waves in atoms

Neils Bohr suggested that the discrete energy levels in atoms are due to the electrons being allowed to have discrete orbits around the atom, rather like Figure 21.7. The changes in energy correspond to changes in orbit.

Erwin Schrödinger suggested that the energy levels in atoms were due to electrons behaving like stationary waves in the atom, with a profile like the waves in Figure 21.8. Only certain types of wave fit the atom, and these correspond to the fixed energy states.

The electromagnetic spectrum

The electromagnetic spectrum comprises a wide range of electromagnetic waves (Table 22.1). All the waves consist of transverse oscillating electric and magnetic waves. Electromagnetic waves can be polarised. They all travel at the same speed of 3.00×10^8 m s^{-1} in a vacuum.

Table 22.1   *The electromagnetic spectrum*

Type	Frequency	Wavelength	How made	Uses	Photon energy
long-wave radio	~250 kHz	~1200 m	oscillating currents in aerials	radio	~10^{-28} J
medium-wave radio	~1000 kHz (1 MHz)	~300 m	oscillating currents in aerials	radio	~10^{-27} J
short-wave radio	~10 MHz	~30 m	oscillating currents in aerials	radio	~10^{-26} J
VHF	~100 MHz	~3 m	oscillating currents in aerials	radio	~10^{-25} J
UHF	~400 MHz	~1 m	oscillating currents in aerials	television	~10^{-25} J
microwaves	~2.5 GHz	~10 cm	directly produced in waveguides	radar, cooking, communicating	~10^{-24} J
infra-red	~10^{14} Hz	~1 μm (> 700 nm)	hot bodies, LEDs	night-sights, heating, short-distance communication	~10^{-19} J ~1 eV
visible	~5×10^{14} Hz	700 – 400 nm	very hot bodies, LEDs	seeing, etc	2 – 3 eV
ultra-violet	>7.5×10^{14} Hz	<400 nm	extremely hot bodies, sparks, discharge tubes	sun-tanning, detecting invisible marking, sterilising	> 3 eV
X-rays	~10^{18} Hz	~10^{-10} m	stopping fast electrons	X-raying people and materials	~10 000 eV
gamma rays (overlap with X-rays)	~10^{20} Hz	~10^{-12} m	nuclear decay	X-raying thick objects, killing cancerous cells, sterilising	~1 MeV ~10^{-13} J
cosmic rays	very high	very short	from distant parts of the Universe	just cause a hazard	up to tens of joules

and therefore

$$v^2 = \frac{2QV}{m} \qquad \text{so} \qquad v = \sqrt{\frac{2QV}{m}}$$

For an electron, $m = 9.1 \times 10^{-31}$ kg and $Q = 1.6 \times 10^{-19}$ C. The electrons in Figure 23.6 were accelerated through 5000 V. Therefore

$$v = \sqrt{\frac{2 \times 1.6 \times 10^{-19}\,\text{C} \times 5000\,\text{V}}{9.1 \times 10^{-31}\,\text{kg}}}$$

$$= 4.2 \times 10^7\,\text{m s}^{-1}$$

and momentum $= mv$ is given as

$$mv = 9.1 \times 10^{-31}\,\text{kg} \times 4.2 \times 10^7\,\text{m s}^{-1} = 3.8 \times 10^{-23}\,\text{kg m s}^{-1}$$

De Broglie's equation states that for a particle with this momentum

$$\lambda = \frac{h}{p} = \frac{6.6 \times 10^{-34}\,\text{J s}}{3.8 \times 10^{-23}\,\text{kg m s}^{-1}} = 1.7 \times 10^{-11}\,\text{m}$$

which is in close agreement with the measurement above.

If you increase the accelerating voltage, the momentum of the electrons increases, so the wavelength decreases. The width of the fringes decreases and the rings close up on the screen.

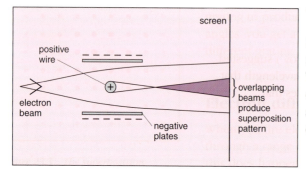

Figure 23.7   *Two-slit superposition with electrons*

Two-slit diffraction with electrons

Figure 23.7 shows how you can demonstrate two-slit superposition with electrons. This arrangement is just like Young's slits for light, but the superposition is harder to demonstrate. The wavelength even for very fast electrons is small, so the slits need to be very close for fringes to be observable.

All matter has wave properties

De Broglie's theory about the wave properties of particles applies to all particles, even large ones like people. The theory suggests that particles are indeed separate particles and that the wave that is associated with them (called their *wave function*) describes the probability of finding the particle in a particular place. Practice question 23.5 invites you to calculate the wavelength of a creeping bacterium. Even for that small particle, the wavelength is so much smaller than the length of the bacterium that the wave effects are simply not observable. Wave properties are only significant for particles of the size of an atom or smaller.

Our universe

Is there anybody out there? Our universe is vast.

Star spectra

Figure 24.1 Stars differ in colour – you can use a diffraction grating to analyse light from stars

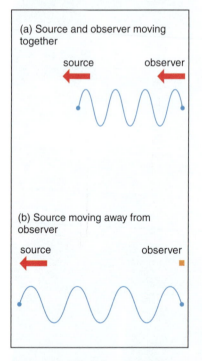

Figure 24.2 Waves from a receding source appear to have a longer wavelength

Star spectra

If you look carefully at the stars in the sky, you can see that some stars are distinctly different in colour from others (Figure 24.1). You can use a diffraction grating, as shown in Chapter 21, to observe the line emission spectra of various elements. You can use a grating to analyse light from individual stars and compare this with light from known elements.

As Figure 21.6 shows, hot gases emit different spectra. These **emission spectra** are characteristic of the different elements that they contain. Sometimes certain frequencies are missing from a spectrum. This can be equally revealing. For instance, sunlight does not produce a continuous visible spectrum but one with some dark lines across it. This is an **absorption spectrum**. Two of these occur in the yellow and denote the absorption of these two frequencies by sodium in the Sun's atmosphere. The presence of absorption lines also confirms that the majority of star matter consists of hydrogen and helium.

Star movements

If a light source and an observer move with the same velocity, the wavelength of the light leaving the source has the same wavelength when it arrives at the observer. But if the source is moving away, then each wave is slightly longer when it reaches the observer, because the source moves slightly by the time the next wave is emitted. The light appears to be shifted towards a longer (redder) wavelength (Figure 24.2).

Red-shift occurs as a result of the source of light moving away from the observer. The faster the speed of the source, the greater the red-shift. If the source of light is moving towards the observer, wavelength decreases, frequency increases, and **blue-shift** occurs. This phenomenon is the **Doppler effect**. The change in both frequency Δf and wavelength $\Delta \lambda$ is related to the speed v at which the star is moving by the equation:

$$\frac{\Delta f}{f} = \frac{\Delta \lambda}{\lambda} = \frac{v}{c}$$

where c is the speed of light.

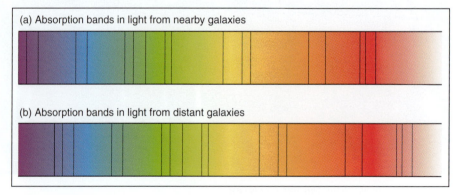

Figure 24.3 Red-shift of light from distant galaxies

Absorption bands in the spectra of distant galaxies appear to be shifted towards the red end of the spectrum (Figure 24.3). This suggests that distant galaxies are moving away from us and that the universe is expanding.

Hubble's law

Astronomers measured the red shift of a large number of galaxies. This shows that more distant galaxies are moving away from us (receding) faster than closer galaxies. Edwin Hubble discovered that the recession velocity is directly proportional to the galaxy's distance from us.

The equation of the line in Figure 24.4 relates the recession velocity v to the distance d:

$$v = Hd$$

where H is the *Hubble constant*. Unfortunately, the value of the Hubble constant is rather uncertain, reflecting the uncertainty in the measurements of astronomical distances. Its value is $(2 \pm 1) \times 10^{-18}\,\text{s}^{-1}$.

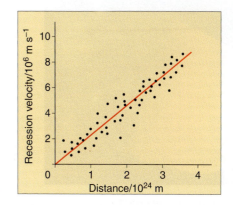

Figure 24.4 Data illustrating Hubble's law

As astronomical distances are so great, they are often measured in light years (ly). A light year is the distance travelled by light in one year and equals 9.46×10^{15} m.

The origin of the universe

All distant galaxies seem to be travelling away from us. If this process has been continuing for some time, the galaxies must once have been much closer, and matter must once have been packed together into a very small volume.

Dividing the distance of a galaxy from us by its speed of travel away from us gives an estimate for the time taken for the expansion.

Assuming that all galaxies are moving away from the same starting point (the Big Bang) and that their individual speeds have not varied, we get:

$$\text{age of universe} = \frac{\text{distance travelled}}{\text{speed}} = \frac{d}{v} = \frac{d}{Hd} = \frac{1}{H}$$

Using the above value for H, we obtain

$$\text{age of universe} = \frac{1}{2 \times 10^{-18}\,\text{s}^{-1}} = 5 \times 10^{17}\,\text{s} \approx 10^{10}\,\text{years}$$

25 | The Big Bang and the Big Crunch

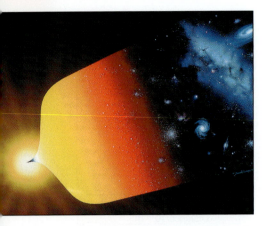

Figure 25.1 An artist's impression of the history of the universe

You read in Chapter 24 how observations of red-shift lead to the theory that the universe is expanding. If we extend this model back in time it gives the idea of the universe starting with a '**Big Bang**' about 15×10^9 years ago. The universe we now live in appears to have expanded from this Big Bang.

All the observations of gravitational forces show that matter attracts other matter. If the universe has expanded from a small volume, gravitational forces will tend to pull it back together. The expanding parts do work against these gravitational forces. Their gravitational potential energy increases but at the same time their kinetic energy, including the random kinetic energy of the parts, decreases.

This leads to the theory that at the time of the Big Bang, the universe was very hot as well as dense. Now it has expanded, the temperature of the universe has cooled down to an average temperature of about 3 K. Figure 25.1 shows an artist's impression of the history of the universe.

The end of the universe

The masses of the galaxies attract each other gravitationally, so the rate of expansion of the universe is probably slowly decreasing. If there is enough mass in the universe, the universe is **closed**: one day it will stop and begin to collapse back towards a single point. If the mass is insufficient for that to happen, the universe is **open**: it will go on expanding for ever but at a slower and slower rate (Figure 25.2).

However, in addition to all the matter in the universe, energy also has mass. So, strictly speaking, whether the universe is closed or open depends on the average mass–energy density of the universe.

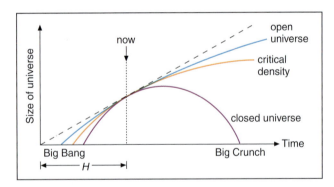

Figure 25.2 Depending on its density, the universe may expand forever, or contract back to a 'Big Crunch'

For collapse to occur, the average density of the universe would have to be about 1×10^{-26} kg m^{-3}, or about six atoms per metre cubed of space. This is called the **critical density** – the boundary between an open and a closed universe. Visible matter accounts for only 2–3% of this critical value. However, the observed movements of galaxies cannot be explained simply in terms of the gravitational attractions between the masses of the visible matter that they contain. The forces seem to be much stronger forces, suggesting that there is a lot of dark matter (not stars) exerting gravitational attraction and placing the average density of the universe much closer to the critical density.

Practice questions

Chapter 1

1.1 Why is it impossible for an object to move with a constant velocity along a circular path?

1.2 How is it possible for an accelerating object to be moving at a constant speed?

1.3 An object is moving round a circular path at a constant speed. Explain why the object must be accelerating towards the centre of the circular path.

1.4 State the origins of the centripetal forces that maintain the following circular motions: (a) the Moon's orbit of the Earth, (b) you riding on a roundabout, (c) a car driving round a bend, (d) you riding on a swing.

1.5 Explain why no work is done by a centripetal force.

Chapter 2

2.1 Show that the equation for centripetal force, $F = mv^2/r$, is homogeneous with respect to its units.

2.2 A 0.5 kg stone on the end of a length of string is whirled round in a horizontal circle, radius 2 m, on a level frictionless table. The stone moves at a constant speed of 7 m s^{-1}. Calculate: (a) the stone's acceleration, stating its direction, (b) the tension in the string, (c) the work done on the stone by the tension during 10 revolutions, explaining your answer.

2.3 Define period, frequency and angular speed, and state the unit in which each is measured.

2.4 A car travels around a circular bend of radius 1 km at a constant speed of 108 km h^{-1}. Calculate its angular speed in rad s^{-1} and its acceleration.

2.5 The diagram shows the starting position of the tape in a 'C90' audiocassette.

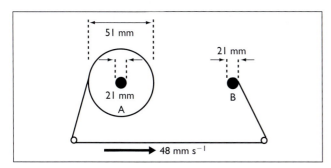

The empty spool (B) has a diameter of 21 mm while that of the full spool (A) is 51 mm. The tape travels from spool A to spool B at a constant speed of 48 mm s^{-1}, taking exactly 45 minutes to empty spool A. Calculate: (a) the total length of tape in the cassette, (b) the initial angular speeds of the two spools. (c) How will the motion of the two spools vary as the tape is being played?

Chapter 3

3.1 A mass attached to the end of a string follows a vertical circular path at a constant speed. The graph shows how the tension in the string varies with time.

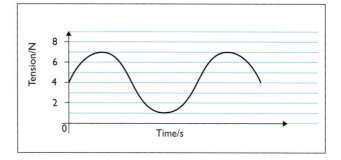

Explain why the tension varies in this way.

3.2 Given that the radius of the vertical circular path in the previous question is 70 cm, use information from the graph to find: (a) the weight of the attached mass, (b) its mass, (c) the speed at which it is moving.

3.3 A fishing line breaks when the force acting on it exceeds 15 N. It is used to provide the force needed to keep a body that weighs 6 N moving around a vertical circular path. Above a certain speed, the fishing line breaks. Whereabouts on its path will the body be when the fishing line breaks? Calculate the maximum speed at which the body could follow a vertical circular path of radius 30 cm without the line breaking.

3.4 What is the difference between weightlessness and apparent weightlessness?

3.5 A conical pendulum consists of a 510 g bob on a 60 cm string. The bob is rotated in a horizontal circle with the string at 30° to the vertical. Draw a free-body force diagram for the bob. Calculate: (a) the tension in the string, (b) the speed of the bob, (c) the period of its motion.

PRACTICE QUESTIONS

Chapter 4

4.1 Explain why a pendulum can be used for timekeeping. Many clocks do not contain a pendulum. State two possible mechanisms on which their timekeeping might be based.

4.2 What is meant by equilibrium position, displacement and amplitude?

4.3 Suggest how a time trace could be obtained for a glider moving between buffers on an air track. Sketch the resulting time trace.

4.4 The lowest note that a clarinet can play has a frequency of 160 Hz. Calculate its wavelength.

4.5 The graph shows how the displacement of an oscillating body varies with time.

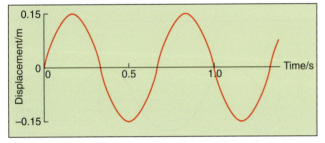

State the amplitude of these oscillations. Determine their frequency. Copy the graph and mark all the points where the body has zero speed.

Chapter 5

5.1 Define simple harmonic motion.

5.2 Sketch a graph to show how force varies with displacement for a simple harmonic oscillator.

5.3 A particle moves with simple harmonic motion between the points A and C.

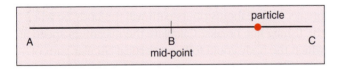

At which point (or points) will the particle have (a) zero velocity, (b) zero acceleration, (c) maximum velocity to the right, (d) maximum acceleration to the left, (e) maximum kinetic energy.

5.4 Sketch graphs to show how displacement, velocity and acceleration vary with time for the same two cycles of a simple harmonic oscillator.

5.5 A light spring hangs vertically from a firm support. A mass of 800 g is attached to the lower end of the spring. It produces a static extension of 4.0 cm. Calculate: (i)

the spring constant k of the spring, (ii) the angular speed ω of the mass when it performs small vertical oscillations, (iii) the period T of these oscillations.

Chapter 6

6.1 An oscillator moves in such a way that its displacement x, measured in centimetres, varies with time t according to the equation $x = 8\cos(4\pi t)$. What are its amplitude, angular speed, period and frequency?

6.2 An oscillator has a period of 0.5 s. Sketch three 'vertically aligned' graphs to show how its displacement, velocity and acceleration vary with time during the initial two seconds of its motion. Calculate the oscillator's maximum velocity and its maximum acceleration.

6.3 An oscillator has a period of π s and an amplitude of 12 cm. Calculate its angular speed. Write down the equations describing how its displacement, velocity and acceleration vary with time.

6.4 Explain, with reference to the phase angles involved, the terms *in phase* and *in antiphase*. If two oscillators always remain in antiphase, what does this tell you about their frequencies?

6.5 Two runners are taking part in a race around a circular track. They are $\dfrac{\pi}{2}$ out of phase. (a) Illustrate this phase difference with a sketch of the track. (b) Sketch two more diagrams to illustrate phase differences of π and 2π.

Chapter 7

7.1 Show that the equation for the period of a mass oscillating on the end of a spring, $T = 2\pi\sqrt{\dfrac{m}{k}}$, is homogeneous with respect to its units.

7.2 A mass on the end of a spring oscillates with a period of 2.4 s. What would be the period if the same mass oscillated on the end of four springs, all identical to the first one used, connected (a) in series, (b) in parallel?

7.3 A simple pendulum has a length of 35 cm. Calculate its period.

7.4 Describe how you would use the oscillations of a pendulum to obtain an accurate value for the acceleration of gravity.

7.5 Explain in detail why a clock can be designed around a system executing simple harmonic motion.

Chapter 8

8.1 Describe in words the changes of energy that occur when a trolley oscillates between springs on a bench. Eventually the trolley stops oscillating. Describe what has happened to the energy.

8.2 As a pendulum oscillates with simple harmonic motion, its energy interchanges between the forms of kinetic and gravitational potential. Sketch a graph showing how, for two complete cycles, its kinetic energy, gravitational potential energy and total energy vary with (a) time, (b) displacement.

8.3 A mass of 600 g is restrained by a horizontal spring system of spring constant 250 N m^{-1}. The mass is displaced 8 cm from the equilibrium position. Calculate the potential energy given to the system. The mass is released. State the kinetic energy of the mass as it passes through its equilibrium position and calculate its speed at this point. Use this speed to estimate the period of the motion. Compare this with the value calculated from $T = 2\pi\sqrt{\dfrac{m}{k}}$. Comment on any difference.

8.4 Describe the energy changes that occur when a mass oscillating on the end of a vertical spring moves from the lowest position through the middle to the highest position of its motion. Assume that the spring remains under tension all the time.

8.5 A light spring is suspended from a rigid support and its free end carries a mass of 0.40 kg, which produces a static extension of 0.06 m in the spring. The mass is pulled down a further 0.06 m and released. The mass then oscillates with simple harmonic motion. By considering all the energy changes involved, calculate the kinetic energy of the mass as it passes through the mid-point of its motion.

Chapter 9

9.1 What is the natural frequency of an oscillator?

9.2 The natural frequency f of a pendulum of mass m and length L is given by the equation $f = \dfrac{1}{2\pi} \times \sqrt{\dfrac{g}{L}}$. Show that this equation is homogeneous with respect to its units. For what value of L would the pendulum have a natural frequency of 1.0 Hz?

9.3 Describe how an oscillator behaves when forced to vibrate at different frequencies ranging from below to above its natural frequency. Explain the term resonance.

9.4 A car's rear-view mirror is on a metal support attached to the roof. The driver notices that the view in the mirror, which is normally sharp, blurs whenever the engine rotates at a certain frequency. Explain these observations and suggest how the problem may be overcome.

9.5 The drums of an automatic washing machine are attached to the casing by strong springs.

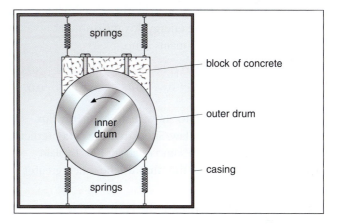

The inner drum rotates within the outer drum at variable speeds according to the washing programme. The total mass of the two drums is 25 kg. A block of concrete of mass 35 kg is added to the outer drum. The natural period of oscillation of the system is 2 s. Calculate the effective spring constant of this mass–spring system. When the washing machine enters the spin part of its programme, the inner drum starts from rest, building up rotational speed gradually. As the speed increases, the system is observed to oscillate with increasing amplitude, reaching a maximum value of 3 cm, before decreasing again at higher speeds. Why does the system oscillate when the inner drum is rotated? How many revolutions per minute will the inner drum be making when the maximum amplitude of oscillation is observed? Explain your answer. Sketch a graph showing how the amplitude of the system varies with the frequency of rotation of the inner drum. State and explain one effect on the oscillations of running the machine without the block of concrete fixed to the outer drum.

Chapter 10

10.1 Describe how a continuous plane (straight) water wave can be produced in the laboratory. How could the frequency of the water wave be changed? How would you measure the wavelength of the water wave and its wave speed?

10.2 What is a progressive wave? How does the amplitude of a progressive wave vary with distance from its source? Give two reasons for this variation.

PRACTICE QUESTIONS

10.3 What is energy flux? The inverse square law can be applied to the energy flux of electromagnetic waves produced by a point source. Explain what is meant by inverse square law. Explain how such a law automatically arises from uniform energy propagation from a point source, stating the principle that has to be applied.

10.4 The table gives the energy flux measured at various distances from a mains lamp.

distance/m	0.50	1.00	1.50	2.00	2.50	3.00
energy flux/W m^{-2}	2.56	0.64	0.28	0.16	0.10	0.07

Use a graphical method to show that the energy flux obeys an inverse square law. Show that the data predicts a power output from the lamp of about 8 W. Explain why the actual power of the lamp will be much greater than this.

10.5 Solar radiation arrives at the Earth's orbit at the rate of 1.4 kW m^{-2}. If the average radius of the Earth's orbit around the Sun is 1.49×10^{11} m, calculate the power output of the Sun.

Chapter 11

11.1 Sound waves are longitudinal while those on the surface of water are mainly transverse. In what ways do these two types of wave differ?

11.2 How would you demonstrate that sound waves moving through air are longitudinal?

11.3 Explain the difference between an unpolarised wave and a plane polarised wave. Why can visible light and microwaves both be polarised while sound waves cannot?

11.4 A grille of metal rods is placed between a microwave transmitter and receiver. Describe and explain the changes in the received amplitude as the grille is rotated about a horizontal axis.

11.5 Describe how you would test whether or not reflected visible light is plane polarised.

Chapter 12

12.1 A hiker uses her binoculars to watch a distant woodcutter at work. She notices that there is a delay of 0.8 s between seeing and hearing the cutting of the axe. Explain this delay. The speed of sound is 330 m s^{-1}. How far away is the woodcutter? After a further 2.4 s delay, the hiker also hears a second sound, an echo from a hill directly behind the woodcutter. How far behind the woodcutter is the hill?

12.2 An oscilloscope is used to time light pulses travelling from a transmitter to a mirror and back to an adjacent receiver. The diagram shows the trace on the oscilloscope's screen corresponding to the reflected pulse with the mirror in two positions.

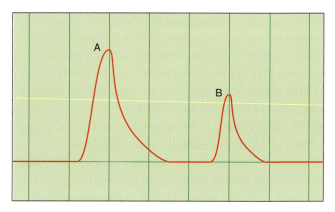

Which trace corresponds to the mirror being further away? Give two reasons for your choice. Given that each horizontal division on the screen corresponds to a time of 20 ns and that the speed of light is 3×10^8 m s^{-1}, calculate how far the mirror was moved between the two positions.

12.3 What is meant by wavefront and wavelength?

12.4 The diagram shows part of a transverse wave travelling from left to right at one instant in time.

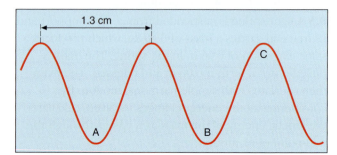

The frequency of its source is 20 Hz. A and B are molecules that at this instant are in neighbouring troughs, 1.3 cm apart. C is on a crest. Calculate the speed at which the wave is travelling. Copy the sketch and add to it the new position of the wave after a further 12.5 ms. Mark the new positions of the molecules A, B and C. Sketch a graph showing how the displacement of molecule A varies with time for 100 ms from the instant shown in the above sketch. Which of the three molecules are vibrating 'in phase' and which are 'in antiphase'?

12.5 A typical value for the speed of sound in air is 330 m s^{-1}. A tuning fork emits a continuous burst of waves that have a wavelength of 8.5 cm. The tuning fork vibrates for a total time of 6.5 s. Calculate (a) the frequency of the emitted sound, (b) the number of complete waves emitted during 20 ms, (c) the total length of the wavetrain emitted.

Chapter 13

13.1 What is diffraction? Explain why it is possible to produce a narrower beam of light than of microwaves.

13.2 When passing through the same narrow slit, red light is found to diffract more than blue light. What can you deduce from this observation? Explain whether bass (low frequency) sounds or treble (high frequency) sounds will diffract the most when passing through the same opening.

13.3 Sketch the diffraction pattern produced when laser light passes through a narrow slit. Compare the widths of the fringes observed.

13.4 Describe, with the aid of a clearly labelled diagram, how you would demonstrate for microwaves that the angle of incidence is equal to the angle of reflection. Explain why it is more difficult to test the laws of reflection for microwaves than for visible light. How would you test the laws of reflection using sound waves?

13.5 A large tray holds water that is deep to the left of the straight line AB and shallow to its right.

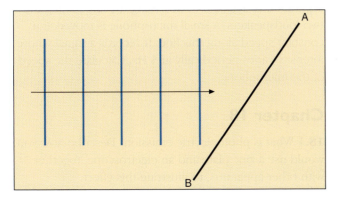

The deep water wave speed is 28 cm s^{-1}, which reduces to 21 cm s^{-1} in the shallow water. Copy and complete the diagram to show the wavefronts as they pass over the boundary and into the shallow region.

Chapter 14

14.1 Describe how you would demonstrate that when two pulses travelling in opposite directions on a spring meet, they pass straight through each other.

14.2 The diagram shows two wave pulses of equal length approaching each other on a spring.

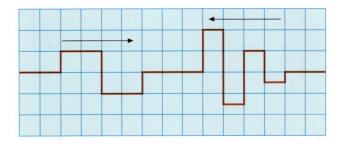

Sketch the appearance of the spring at the instant when the two pulses fully overlap.

14.3 State the principle of superposition. Explain in terms of the phase of the waves involved the existence of constructive and destructive superposition.

14.4 What are coherent sources? Explain why two sources are unable to produce a stable superposition pattern unless they have the same frequency.

14.5 Two coherent sources are found to produce a stable and well-defined superposition pattern. State and explain the additional fact that this tells you about the two sources.

Chapter 15

15.1 State the path differences that correspond to the following phase differences, all measured in radians: 0, $\pi/2$, π, $3\pi/2$, 2π

15.2 Two loudspeakers are connected to a 2 kHz supply and placed 2 m apart at the front of a laboratory. Students walk around the room and find that there is a minimum intensity along a line that is equidistant from both loudspeakers. What does this tell you about the way in which the two loudspeakers are connected? How must the apparatus be adjusted to produce a central maximum along this same line?

15.3 Explain how a two-source superposition experiment can be carried out using just a single source.

15.4 The diagram shows the position P of the third maximum to one side of the central maximum of the superposition pattern of the two coherent sources, S_1 and S_2.

PRACTICE QUESTIONS

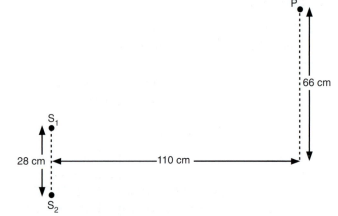

Calculate the wavelength of the sources.

15.5 The diagram shows microwaves arriving at a receiver R having travelled along two different paths from the same transmitter T.

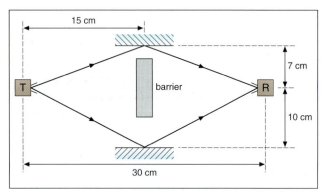

Explain whether or not these waves will be coherent. The wavelength of the microwaves is 2.9 cm. By calculating the path difference, deduce whether the receiver shown is detecting a maximum or a minimum intensity.

Chapter 16

16.1 Explain why a small, single slit is placed in front of a filament lamp when it is used to produce superposition patterns. Why is a single slit not necessary when using a laser?

16.2 What is meant by monochromatic? Draw a diagram of the apparatus you would use to produce a two-slit superposition pattern for monochromatic visible light. Mark any relevant dimensions on your diagram.

16.3 Sketch the superposition pattern produced when monochromatic light illuminates a double slit. Compare this to the pattern produced using a single slit. What is the effect of using white light?

16.4 A laser produces red light with a wavelength of 680 nm. Using this light, a superposition pattern with a fringe width of 3.2 mm is produced on a screen 1.8 m from a double slit. Calculate the slit separation of the double slit.

16.5 A light source of wavelength 550 nm illuminates two slits 0.25 mm apart. What will be the fringe width on a screen 2 m from the slits?

Chapter 17

17.1 What is a stationary wave? What conditions are required for the production of a stationary wave?

17.2 What are nodes and antinodes? Explain how nodes and antinodes are formed.

17.3 What is fundamental frequency? The fundamental frequency f of a stationary wave on a stretched rubber cord of length l is given by $f = \dfrac{1}{2l}\sqrt{\dfrac{T}{\mu}}$ where T is the tension in the cord and μ is its mass per unit length. Show that this equation is homogeneous with respect to its units. What happens to the fundamental frequency when the tension in the cord is increased from 40 N to 160 N?

17.4 Describe four ways in which progressive and stationary waves are different.

17.5 Two loudspeakers are connected to the same oscillator and placed facing each other several metres apart. The oscillator produces a frequency of 850 Hz. The speed of sound in air is 330 m s^{-1}. Calculate the separation of adjacent nodes along the line joining the two loudspeakers. A small microphone is moved at a constant speed along this line. It records a signal whose intensity varies periodically at 5 Hz. Calculate the speed of the microphone.

Chapter 18

18.1 What is photoelectric emission? Describe how you would use a zinc plate and an electroscope, together with other apparatus, to illustrate this effect.

18.2 What is threshold frequency? A more accurate value for the threshold frequency of zinc is 0.88×10^{15} Hz. Given that the speed of light is 3.0×10^8 m s^{-1}, calculate the threshold wavelength of zinc.

18.3 Explain why, for a given surface, there is a maximum wavelength of incident radiation above which photoelectric emission cannot be observed. The magnitude of this maximum wavelength is found to increase slightly if the temperature of the photoelectric surface is increased. Suggest a reason for this.

18.4 What is a photon? In terms of photons, what is the difference between (a) a bright light source and a dim one of the same frequency, (b) a visible light source and an ultra-violet source of the same intensity?

18.5 What is work function? Explain why photoelectric emission occurs from zinc when illuminated with a weak ultra-violet source, but not when an intense visible source is used?

Chapter 19

19.1 Explain why the fastest moving photoelectrons will be those freed from near the surface of the material.

19.2 Describe how you would measure the kinetic energy with which the fastest photoelectrons are released.

19.3 Show that the relationship:

photon energy = work function + kinetic energy of fastest electrons

leads to the equation:

$$\text{maximum kinetic energy} = \frac{hc}{\lambda} - \phi$$

where $c = 3.00 \times 10^8 \text{ m s}^{-1}$

In an experiment with a vacuum photocell, the maximum kinetic energy of the electrons emitted was measured for different wavelengths of the illuminating radiation. The following results were obtained.

maximum kinetic energy/10^{-19} J	3.26	2.56	1.92	1.25	0.58
incident wavelength/10^{-7} m	3.00	3.33	3.75	4.29	5.00

Use the results to plot a graph of maximum kinetic energy against (1/incident wavelength). Use your graph to determine values for Planck's constant h and the work function of the photoelectric surface ϕ.

19.4 The work function for sodium metal is 3.78×10^{-19} J. A fresh sodium surface is irradiated with ultra-violet light of wavelength 319 nm. Calculate the maximum kinetic energy of the emitted photoelectrons. What will be the value of the stopping voltage for these electrons?

19.5 A metal surface is illuminated with light of wavelength 500 nm. Photoelectrons are liberated with a maximum kinetic energy of 2.4×10^{-19} J. Calculate the wavelength of the incident radiation required to liberate electrons with a maximum kinetic energy of 9.0×10^{-19} J from the same surface.

Chapter 20

20.1 What is an electronvolt? Calculate the energy, both in joules and electronvolts, of a photon of light of wavelength 253 nm.

20.2 A material has a work function of 1.4 eV. Calculate the work function in joules. What is the longest wavelength of incident radiation that can cause photoelectrons to be emitted from its surface?

20.3 Calculate the photon energy of light of wavelength 0.40 µm. A caesium surface (work function 1.94 eV) is illuminated with this light. Calculate the maximum kinetic energy of the emitted electrons. The mass of an electron is 9.1×10^{-31} kg. Calculate the speed of the fastest electrons.

20.4 When photoelectric emission is occurring, what determines (a) the number of photoelectrons released each second, (b) the maximum kinetic energy of the emitted photoelectrons?

20.5 A zinc electrode of work function 0.88 eV is illuminated by light of wavelength 250 nm. Calculate the photon energy of the light in electronvolts and the stopping voltage. Explain why, for a given material, the stopping voltage increases with the frequency of the light used.

Chapter 21

21.1 Explain the terms ionisation energy, ground state and excited state. The ionisation energy for a mercury atom in its ground state is 10.4 eV. When in its first excited state, only 5.5 eV is needed to ionise it. How much energy is needed to raise a mercury atom from its ground state to its first excited state?

21.2 The ionisation energy of sodium is 5.14 eV. Calculate the speed of the slowest electron (electron mass = 9.1×10^{-31} kg) that can ionise a sodium atom by collision.

21.3 Using data from Figure 21.3 on page 44, calculate the energy released when a hydrogen atom passes from its second excited state to its first. What wavelength of light has this photon energy? To which part of the electromagnetic spectrum does this wavelength belong?

21.4 What is an emission spectrum? When light of wavelength λ passes through a diffraction grating that has N slits per metre, the nth-order spectrum is deviated through an angle θ such that $n\lambda = \dfrac{\sin \theta}{N}$.

Show that this equation is homogeneous with respect to its units. Describe how you would use a diffraction grating to observe an emission spectrum.

PRACTICE QUESTIONS

21.5 Compare the models suggested by Bohr and Schrödinger to explain the existence of discrete energy levels in atoms.

Chapter 22

22.1 One type of electromagnetic wave is used for television broadcast. A television aerial is pointed towards a transmitter while being rotated through 90°. Reception is found to fall from a maximum to zero. What two things can you deduce about 'television waves' from this observation?

22.2 The electromagnetic spectrum consists of seven main divisions; three of these are named in the following table:

| radiowaves | (a) | (b) | visible | (c) | (d) | gamma rays |

Give the names of the divisions represented by the letters (a)–(d). What quantity decreases as you move left-to-right across the table? In which division would you find a wavelength of (a) 3 cm, (b) 600 nm?

22.3 The table lists some of the electromagnetic frequencies and wavelengths used in radio and television broadcasting.

Calculate the missing values.

Type	Frequency / Hz	Wavelength / m
a.m. radio	200 kHz	(a)
f.m. radio	(b)	3.24 m
u.h.f. television	0.516 GHz	(c)

22.4 A transmitter for Radio 5 emits a wave of frequency 909 kHz at a power of 12 kW. Calculate the energy of a single photon of this radiation. How many photons leave the transmitter each second? An aerial with an area of 0.01 m^2 is 20 km from the transmitter. Assuming the transmitter acts as a point source, calculate the number of photons incident each second on the aerial.

22.5 X-rays are electromagnetic waves in the frequency range 10^{17} Hz to 10^{21} Hz. Calculate the wavelength range of X-rays. When a fast-moving electron is brought to a halt, X-rays may be produced. If all the kinetic energy of an electron travelling at 2.3×10^7 m s^{-1} is transformed into one photon of X-radiation, calculate the frequency of this X-radiation.

Chapter 23

23.1 Explain, with reference to the behaviour of both photons of light and high-speed electrons, what you understand by the phrase *wave-particle duality*. Give examples of both types of behaviour in each case.

23.2 An electron beam is accelerated from rest through a potential difference of 1500 V. Calculate the final speed, momentum and de Broglie wavelength of one of the accelerated electrons. (The mass of an electron = 9.1×10^{-31} kg and the charge = 1.6×10^{-19} C.)

23.3 A 'grating' is used to diffract an accelerated electron beam of wavelength 3.2×10^{-11} m. Calculate the spacing of the slits necessary to give a first-order 'electron' image at an angle of 12° to the central maximum. How might such a grating spacing be achieved?

23.4 Sketch the image produced on the screen of an electron diffraction tube. Mark on your sketch the places of constructive and destructive interference. The accelerating voltage of the tube is increased. What effect does this have on the speed of the electrons and their wavelength? Describe the effect that it will have on the pattern.

23.5 Calculate the wavelengths of the following: (a) a student (mass 60 kg) moving at 2 m s^{-1}, (b) an extremely thin (7 kg!) and slow student moving at 3 mm s^{-1}, (c) an electron, mass 9.1×10^{-31} kg moving at 6×10^7 m s^{-1}, (d) a bacterium (mass 10^{-15} kg) creeping along at 0.3 μm per second. Use your answers to (a), (b) and (c) to explain why the electron will be diffracted by a carbon lattice of spacing 10^{-9} m, but neither student will be diffracted by a door 0.5 m wide. A bacterium is about 1 μm long. Discuss whether or not diffraction will be noticeable with tiny creatures.

Chapter 24

24.1 Describe the differences between emission and absorption spectra.

24.2 What is the Doppler effect? What evidence, provided by the Doppler effect, suggests that the universe is expanding? The Doppler effect occurs with all types of wave motion. Describe and explain the change in the sound of the siren of a passing ambulance.

24.3 A certain galaxy is thought to be moving away from the Earth at a speed of 1.2×10^7 m s^{-1}. Calculate the apparent wavelength, measured using light from this galaxy, of a spectral line whose normal wavelength is 410.2 nm ($c = 3 \times 10^8$ m s^{-1})

24.4 Use a Hubble constant of 1.7×10^{-18} s^{-1} to estimate how far the galaxy in the previous question is from the Earth. Use the speed of the galaxy to estimate how long it has taken to reach its present distance from the Earth. Explain why both your answers are only estimates.

24.5 Show that a light year is equivalent to about 9.5×10^{15} m. The speed of light is 3×10^8 m s^{-1}.

Chapter 25

25.1 Discuss how the ultimate fate of the universe depends on its average mass–energy density.

25.2 It can be shown that the critical density ρ_o of the universe is related to the Hubble constant H and the universal gravitational constant G by the equation:

$$\rho_o = \frac{3H^2}{8\pi G}$$

Show that this equation is homogeneous with respect to its units. Why is it impossible to calculate a completely accurate value for the critical density using this equation?

25.3 For the universe to be closed, its average density would have to be greater than about 1×10^{-26} kg m^{-3}. Compare this with the average density of the Earth, mass 6×10^{24} kg and radius 6400 km.

25.4 What is the difference between visible and dark matter? Which type contributes most to the total mass of the universe?

25.5 Sir Fred Hoyle, who died in 2002, was a famous physicist. He favoured a *steady state* theory of the universe and coined the term 'Big Bang' to try to ridicule that theory. Some scientists still prefer the steady state theory. Use encyclopaedias or the Internet to find out about both theories. Explain which observations support which theory.

Assessment questions

The following questions have been chosen or modified to be similar in style and format to those which will be set for the A2 assessment tests. These questions meet the requirements of the assessment objectives of the specification.

1 Explain how a body moving at constant speed can be accelerating. **[3]**

The Moon moves in a circular orbit around the Earth. The Earth provides the force which causes the Moon to accelerate. In what direction does this force act? **[1]**

There is a force which forms a Newton's third law pair with this force on the Moon. On what body does this force act and in what direction? **[2]**

(Total 6 marks)
(Edexcel GCE Physics Module Test PHI, June 1995)

2 A satellite orbits the Earth once every 120 minutes. Calculate the satellite's angular speed in rad s^{-1}. **[2]**

The satellite is in a state of free fall. What is meant by the term free fall? How can the height of the satellite stay constant if the satellite is in free fall? **[3]**

(Total 5 marks)
(Edexcel GCE Physics Module Test PHI, June 1996)

3 A tennis ball connected to a long piece of string is swung around in a horizontal circle above the head of a pupil.

The pupil feels that there is a tension in the string and argues that for equilibrium there must be an outward 'centrifugal' force acting on the ball. Criticise his argument and explain why there is a tension in the string. **[5]**

The pupil lets go of the string. Draw a free-body force diagram for the ball at the instant after release. **[1]**

(Total 6 marks)
(Edexcel GCE Physics Unit Test PHY4, January 2002)

4 A stone on a string is whirled in a vertical circle of radius 80 cm at a constant angular speed of

16 rad s^{-1}. Calculate the speed of the stone along its circular path. **[2]**

Calculate its centripetal acceleration when the string is horizontal. **[2]**

Explain why the string is most likely to break when the stone is nearest the ground. **[2]**

(Total 6 marks)
(Edexcel GCE Physics Module Test PHI, January 1997)

5 What is meant by simple harmonic motion? **[2]**

Calculate the length of a simple pendulum with a period of 2.0 s. **[2]**

The graph shows the variation of displacement with time for a particle moving with simple harmonic motion.

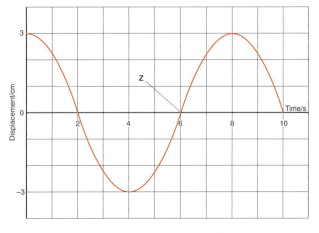

State the amplitude of the oscillation. **[1]**

Estimate the speed of the particle at the point labelled Z. **[2]**

Draw a graph of the variation of velocity with time for this particle over the same period of time. Add scales to both axes. **[2]**

(Total 9 marks)
(Edexcel GCE Physics Module Test PH2, June 1997)

6 The graph shows the variation of acceleration a with displacement x for a body oscillating with simple harmonic motion.

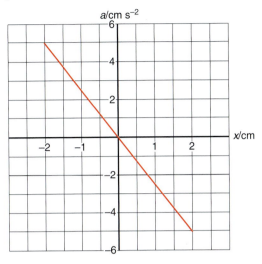

Calculate the period of oscillation of the body. **[3]**

At time $t = 0$ s the body is momentarily at rest. Sketch a graph to show how acceleration of the body varies with time from $t = 0$ s to $t = 12$ s. Add scales to both axes. **[4]**

(Total 7 marks)

(Edexcel GCE Physics Module Test PH2, June 1999)

7 A simple pendulum of length l has a bob of mass m.

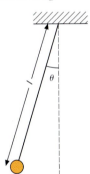

A student studies the variation of its time period T with the angle θ (which is a measure of the amplitude of the motion), the mass m and the length l.

Sketch graphs to show how T varies with θ and how T varies with m. **[2]**

Describe how the student could verify experimentally that $T \propto l$. **[4]**

Below is a graph of $\dfrac{T^2}{4\pi^2}$ against l.

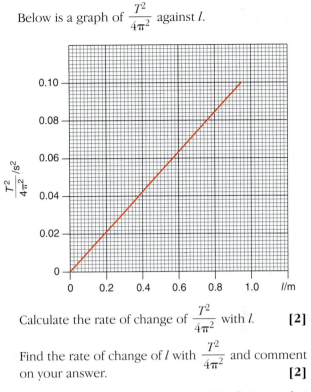

Calculate the rate of change of $\dfrac{T^2}{4\pi^2}$ with l. **[2]**

Find the rate of change of l with $\dfrac{T^2}{4\pi^2}$ and comment on your answer. **[2]**

(Total 10 marks)

(Edexcel GCE Physics Unit Test PHY4, January 2002)

8 Diagram A shows a mass suspended by an elastic cord. The mass is pulled downwards by a small amount and then released so that it performs simple harmonic oscillations of period T. Diagrams B–F show the positions of the mass at various times during a single oscillation.

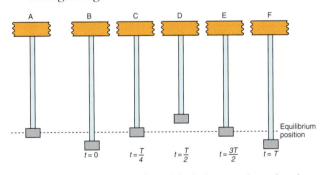

Copy and complete the table below to describe the displacement, acceleration and velocity of the mass at the stages B–F, selecting appropriate symbols from the following list:

maximum and positive → +

maximum and negative → −

zero → 0

Use the convention that downward displacements, accelerations and velocities are positive.

ASSESSMENT QUESTIONS

	Displacement	Acceleration	Velocity
B			
C			
D			
E			
F			

[4]

In the sport of bungee jumping, one end of an elastic rope is attached to a bridge and the other end to a person. The person then jumps from the bridge and performs simple harmonic oscillations on the end of the rope. Some people are bungee jumping from a bridge 50 m above a river. A jumper has a mass of 80 kg and is using an elastic rope of unstretched length 30 m. On the first fall the rope stretches so that at the bottom of the fall the jumper is just a few millimetres above the water. Calculate the decrease in gravitational potential energy of the bungee jumper on the first fall. **[2]**

What has happened to this energy? **[1]**

Calculate the force constant k, the force required to stretch the elastic rope by 1 m. **[3]**

Hence calculate T, the period of oscillation of the bungee jumper. **[2]**

(Total 12 marks)
(Edexcel GCE Physics Module Test PH2, January 1998)

9 A clever method of 'weighing' very small objects, such as tiny carbon particles, is to attach them to a nanotube. The carbon particle is set into vibration.

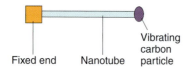

Fixed end   Nanotube   Vibrating carbon particle

In one such experiment, the carbon particle vibrates with maximum amplitude when at a frequency of 970 kHz.

What name is given to the frequency at which an object vibrates with maximum amplitude? **[1]**

This arrangement can be modelled as a mass on a spring. Calculate the mass of the carbon particle, assuming that the spring constant is 0.81 N m^{-1}. **[3]**

What assumption about the motion of this tiny object has been made? **[1]**

(Total 5 marks)
(Edexcel GCE Physics Unit Test PHY4, January 2002)

10 The following describes an experiment which demonstrates forced vibrations.

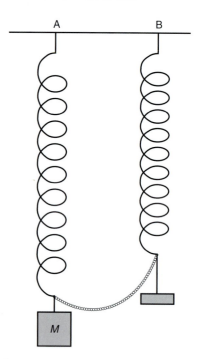

Two identical springs are suspended from a rigid support. Spring A carries a mass M kg while spring B carries a hanger to which slotted masses can be added. The mass of the hanger is much less than M. The springs are linked by a loosely hanging chain as shown.

Mass M is displaced and performs vertical oscillations only. After a few seconds the hanger on spring B is observed to be oscillating vertically with a very small amplitude. The experiment is repeated several times with an extra mass added to the hanger on spring B each time, until the total mass on B is $2M$ kg. Describe and explain the changes to the oscillations of both springs as the mass on B is increased. **[6]**

(Total 6 marks)
(Edexcel GCE Physics Module Test PH2, January 2000)

11 The diagram shows the shape of a wave on a stretched rope at one instant of time. The wave is travelling to the right.

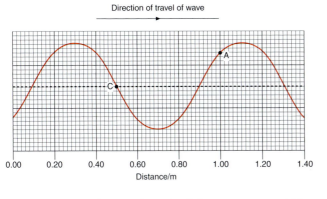

Direction of travel of wave

Distance/m

Determine the wavelength of the wave. **[1]**

On a copy of the diagram:

Mark a point on the rope whose motion is exactly out of phase with the motion at point A. Label this point X.

Mark a point on the rope which is at rest at the instant shown. Label this point Y.

Draw an arrow at point C to show the direction in which the rope at C is moving at the instant shown. **[3]**

The wave speed is 3.2 m s^{-1}. After how long will the rope next appear exactly the same as in the diagram above? **[2]**

(Total 6 marks)
(Edexcel GCE Physics Unit Test PHY4, June 2002)

12 Sound travels by means of longitudinal waves in air and solids. A progressive sound wave of wavelength λ and frequency f passes through a solid from left to right. The diagram X below represents the equilibrium positions of a line of atoms in the solid. Diagram Y represents the positions of the same atoms at a time $t = t_0$.

Explain why the wave is longitudinal. **[1]**

On a copy of diagram Y label (i) two compressions (C), (ii) two rarefactions (R), (iii) the wavelength λ of the wave. **[3]**

The period of the wave is T. Give a relationship between λ, T and the speed of the wave in the solid. **[1]**

(Total 5 marks)
(Edexcel GCE Physics Module Test PH2, June 1996)

13 Explain the term plane polarised wave. **[2]**

Describe an experiment using light or microwaves which tests whether or not the waves are plane polarised. **[2]**

The following statements are all about progressive sound waves:

(i) Since sound waves are longitudinal they cannot be diffracted.

(ii) Sound waves transmit pressure but not energy.

(iii) A sound wave of frequency 436 Hz travelling at 331 m s^{-1} has a wavelength of 75 cm $\pm$ 1 cm.

State whether each statement is true or false. **[3]**

(Total 7 marks)
(Edexcel GCE Physics Module Test PH2, January 1998)

14 The diagram shows wavefronts spreading out from two identical sources, S_1 and S_2.

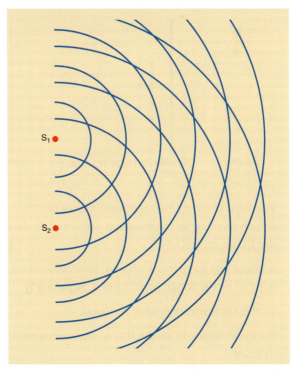

Describe how such a pattern could be produced and observed using a ripple tank. **[5]**

ASSESSMENT QUESTIONS

On a copy of the diagram draw the following:

(i) a line labelled A joining points where the waves from S_1 and S_2 have travelled equal distances,

(ii) a line labelled B joining points where the waves from S_1 have travelled one wavelength further than the waves from S_2,

(iii) a line labelled C joining points where the waves from S_2 have travelled half a wavelength further than the waves from S_1. **[4]**

Complete both of the sentences below by selecting an appropriate term from the following:

     increase   decrease   stay the same

If only the separation of the sources were increased, the angle between lines A and B would ....

If only the wavelength of the waves were increased, the angle between lines A and B would .... **[2]**

(Total 11 marks)

(Edexcel GCE Physics Module Test PH2, June 1999)

15 A laser emits green light of wavelength 540 nm. The beam is directed onto a pair of slits as shown.

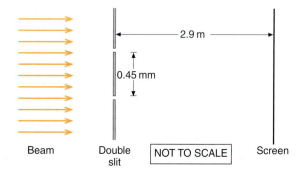

The light from the two slits superposes on the screen forming an interference pattern. Calculate the fringe separation. **[2]**

Without any further calculation, state what would happen to the fringe separation if, **separately**,

(i) the slit separation were reduced,

(ii) the distance from the slits to the screen were increased,

(iii) the laser were replaced with one which emitted red light. **[3]**

Draw the diffraction pattern you would observe if **one** of the slits were covered up. **[3]**

(Total 8 marks)

(Edexcel GCE Physics Unit Test PHY4, June 2002)

16 Under what circumstances could two progressive waves produce a stationary wave? **[2]**

Describe with the aid of a diagram an experiment using microwaves to produce stationary waves. How would you show that a stationary wave had been produced? **[3]**

(Total 5 marks)

(Edexcel GCE Physics Module Test PH2, June 2000)

17 The apparatus shown is used to set up a stationary wave on a stretched string. When the frequency of the vibrator is 60 Hz, resonance occurs and the stationary wave shown is produced.

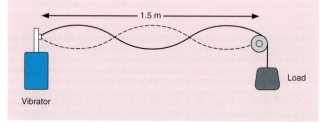

The following statements are all about this apparatus:

(i) When the frequency of the vibrator is 160 Hz there will be eight loops on the string.

(ii) The speed of the wave is 30 m s^{-1}.

(iii) Resonance will also occur when the frequency is 40 Hz.

State and explain whether each statement is true or false. **[6]**

(Total 6 marks)

(Edexcel GCE Physics Module Test PH2, June 1996)

18 The diagram shows apparatus which can be used to demonstrate the photoelectric effect.

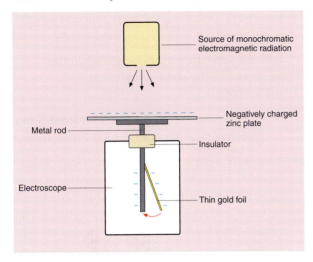

The deflection of the thin gold foil is a measure of the charge stored on the zinc plate. When ultraviolet light is directed towards the zinc plate, the thin gold foil gradually returns to the vertical. When red light is used the thin gold foil stays in the position shown. How does the particle theory of light explain these observations? **[4]**

What would be observed in each case if electromagnetic radiation of greater intensity were used? **[2]**

What would be observed if the zinc plate and electroscope were positively charged? Explain your answer. **[2]**

(Total 8 marks)
(Edexcel GCE Physics Module Test PH2, June 2000)

19 The graph shows how the intensity of light from a light-emitting diode (LED) varies with distance from the LED.

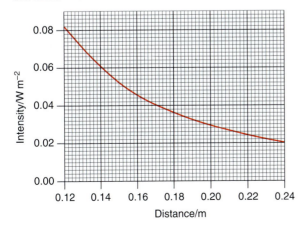

Use data from the graph to show that the intensity obeys an inverse square law.

What does this suggest about the amount of light absorbed by the air? **[3]**

The light from the LED has a wavelength of 620 nm. Show that the energy of a photon of this light is approximately 3×10^{-19} J. **[2]**

A student observes the LED from a distance of 0.20 m. The pupil of her eye has a diameter of 6.0 mm. Calculate the number of photons which enter her eye per second. **[4]**

Explain in terms of photons why the light intensity decreases with increasing distance from the LED. **[1]**

(Total 10 marks)
(Edexcel GCE Physics Unit Test PHY4, June 2002)

20 The diagram shows monochromatic radiation falling on a photocell connected to a circuit.

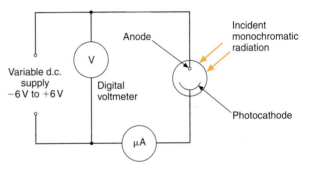

The incident radiation has a wavelength of 215 nm. The metal surface of the photocathode has a work function of 2.26 eV.

Calculate the energy, in eV, of a photon of the incident radiation. **[4]**

What is the maximum kinetic energy, in eV, of the emitted electrons? **[1]**

Write down the value of the stopping potential. **[1]**

(Total 6 marks)
(Edexcel GCE Physics Unit Test PHY4, January 2002 [part question])

ASSESSMENT QUESTIONS

21 Monochromatic light of constant intensity falls on a photocell. The graph shows how the current in the photocell varies with the potential difference applied across it.

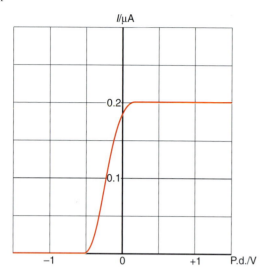

The frequency of the incident light is 6.0×10^{14} Hz. Use the graph to estimate the work function of the metal which forms the cathode of the photocell. [3]

Copy the graph and add to it:

(i) the graph obtained when only the intensity of the light is increased. Label this graph A.

(ii) the graph obtained when only the frequency of the light is increased. Label this graph B. [4]

(Total 7 marks)
(Edexcel GCE Physics Module Test PH2, January 2000)

22 The graph shows how the maximum kinetic energy T of photoelectrons emitted from the surface of a sodium metal varies with the frequency f of the incident electromagnetic radiation.

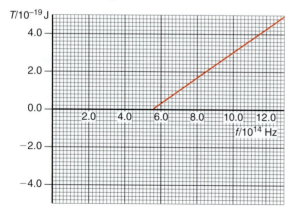

Use the graph to find a value for the Planck constant. [3]

Use the graph to find the work function φ of sodium metal. [2]

Calculate the stopping potential when the frequency of the incident radiation is 9.0×10^{14} Hz. [3]

(Total 8 marks)
(Edexcel GCE Physics Unit Test PHY4, June 2002)

23 A muon is a particle which has the same charge as an electron but its mass is 207 times the mass of an electron. An unusual atom similar to hydrogen has been created, consisting of a muon orbiting a single proton. An energy level diagram for this atom is shown.

0 eV	
−312 eV	
−703 eV	
−2810 eV	Ground state

State the ionisation energy of this atom. [1]

Calculate the maximum possible wavelength of a photon which, when absorbed, would be able to ionise this atom. To which part of the electromagnetic spectrum does this photon belong? [4]

Calculate the de Broglie wavelength of a muon travelling at 11% of the speed of light. [3]

(Total 8 marks)
(Edexcel GCE Physics Module Test PH2, June 1999)

24 Listed below are four types of wave:

microwave   sound   ultraviolet   infrared

From this list, choose the wave which matches each description below. You may choose a type of wave once, more than once or not at all.

A wave capable of causing photoelectric emission of electrons **[1]**

A wave whose vibrations are parallel to the direction of propagation of the wave **[1]**

A transverse wave of wavelength 5×10^{-6} m **[1]**

The wave of highest frequency **[1]**

(Total 4 marks)
(Edexcel GCE Physics Unit Test PHY4, January 2002)

25 Explain what is meant by the term wave-particle duality. **[3]**

Calculate the de Broglie wavelength of a snooker ball of mass 0.06 kg travelling at a speed of 2 m s^{-1}. Comment on your answer. **[2,1]**

(Total 6 marks)
(Edexcel GCE Physics Module Test PH2, January 1997)

26 The electron in a hydrogen atom can be described by a stationary wave which is confined within the atom. This means that its de Broglie wavelength must be similar to the size of the atom, of the order of 10^{-10} m.

Calculate the speed of an electron whose de Broglie wavelength is 1.0×10^{-10} m. **[3]**

Calculate the kinetic energy of this electron, in electronvolts. **[2]**

When β radiation was first discovered, it was suggested that the atomic nucleus must contain electrons. However, it was soon realised that this was impossible because such electrons would possess far too much energy to be bound within the nucleus. Using the ideas of the earlier parts of this question, suggest why an electron confined within the nucleus would have a very high energy. **[2]**

(Total 7 marks)
(Edexcel GCE Physics Unit Test PHY4, June 2002)

27 Describe briefly how the component of a star's velocity towards or away from the Earth may be found. **[3]**

State Hubble's law and illustrate your statement with a sketch graph. Use your graph to explain the meaning of the Hubble constant. **[3]**

Assuming a value of 3×10^{-18} s^{-1} for the Hubble constant, find the distance from the Earth of a galaxy for which the red shift for a particular spectral line is a tenth of the wavelength of the same spectral line observed from a stationary source. **[3]**

(Total 9 marks)
(Edexcel GCE Physics Module Test PH3, January 2000)

28 The Doppler shift may be used in the study of distant galaxies. Explain what is meant by a Doppler shift and how it is used to deduce the motion of distant galaxies. You may be awarded a mark for the clarity of your answer. **[5]**

The graph shows the variation of the size S of an open universe against time t.

On the same axes, sketch a second graph showing how S varies with t for a closed universe. **[1]**

It can be shown that the universe is closed if its density exceeds a critical value ρ. This is determined from the Hubble constant H using
$\rho = kH^2$ where k is a known constant.

Outline the experimental difficulties in determining ρ accurately. **[3]**

(Total 9 marks)
(Edexcel GCE Physics Unit Test PHY4, January 2002)

Things you need to know

Chapter 1 Going round in circles

uniform circular motion: travelling in a circular path with constant speed

centripetal acceleration: acceleration of a body following a circular path; directed towards centre of circle

centripetal force: resultant force that has to acts towards the centre of a circle to make a body follow a circular path

Chapter 2 Circular motion calculations

period: time for one complete rotation

frequency: number of complete rotations each second

radian: central angle where arc length = radius

angular speed: rate at which central angle changes each second

Chapter 3 Circular motion under gravity

weightless: having no weight

apparent weightlessness: situation that occurs when a body is in free fall; the only force on the body is its weight.

Chapter 4 Periodic motion

periodic motion: movement that repeats itself in a regular manner

oscillation: regular, back-and-forth motion

equilibrium position: where the resultant force on an oscillating body is zero

displacement x: how far, and in what direction, the body is from its equilibrium position

time trace: sketch showing variation of displacement with time

amplitude x_0: maximum displacement

cycle: a complete movement of an oscillating system

period T: time taken to cover one complete oscillation or cycle

frequency: number of complete oscillations each second

Chapter 5 Simple harmonic motion

simple harmonic motion (s.h.m.): oscillatory motion where the period does not depend on the amplitude

sinusoidal: the sine or cosine shape of the time trace associated with s.h.m.

definition of s.h.m.: motion where the acceleration (or force) is directly proportional to the displacement from the equilibrium position and always directed towards that central position

Chapter 6 Oscillations and circular motion

phase angle: fraction of a complete oscillation between the oscillations of one oscillator and another

in phase: in step with each other

out of phase (antiphase): completely out of step with each other

Chapter 9 Mechanical resonance

natural frequency: the frequency at which a free-standing system oscillates after it has been displaced and then released

resonance: the large-amplitude oscillations that arise as a result of an oscillatory system being driven at a frequency equal to its natural frequency

damping: forces acting against oscillatory motion, reducing amplitude

Chapter 10 Travelling waves

electromagnetic wave: a transverse wave of oscillating electric and magnetic fields

travelling wave (or progressive wave): a disturbance that transfers energy

energy flux (or intensity): energy that a wave carries perpendicularly through unit area each second

point source: emits uniformly in all directions

inverse square law: when a quantity decreases in proportion to the square of the increasing distance

Chapter 11 Transverse and longitudinal waves

transverse wave: a wave where the displacements are perpendicular to the direction of propagation

longitudinal wave: a wave where the displacements are parallel to the direction of propagation

plane polarised: vibrations are confined to a single plane perpendicular to the direction of energy propagation

unpolarised: vibrations occur in a large number of planes perpendicular to the direction of energy propagation

Chapter 12 Wave speed, wavelength and frequency

wavefront: line joining all points across adjacent rays that have the same phase

wavelength: the minimum distance between two in-phase points on a wave

wave speed: the rate at which the outline of a wave travels through a medium

Chapter 13 Bending rays

diffraction: the spreading out of a wave as it passes through an aperture

ray: a narrow beam

reflection: where waves hit and rebound from a barrier and remain in the same medium

refraction: the change in direction of a wave as it passes from one medium into another in which it has a different speed

Chapter 14 The principle of superposition

principle of superposition: resultant displacement at any point is equal to the vector sum of the displacements of the individual waves at that point at that instant

superposition pattern: a pattern of large- and small-amplitude waves produced where sets of waves overlap

constructive superposition: combination of in-phase waves to produce a wave of increased amplitude

destructive superposition: combination of out-of-phase waves to produce a wave of reduced amplitude

coherent sources: sources that maintain a constant phase relationship

Chapter 15 Two-source superposition experiments

central maximum: line of constructive superposition along which the path difference is zero

path difference: how much further one route is than another; determines the phase of combining waves

Chapter 16 Superposition of light

subsidiary maxima: maxima either side of the central maximum

Chapter 17 Stationary waves

stationary wave: a disturbance that does not transfer energy although it does have energy associated with it

node: a point on a stationary wave where the displacement is always zero

antinode: a point on a stationary wave that oscillates with the maximum amplitude

fundamental frequency: the lowest frequency at which a stationary wave occurs for a given system

harmonic frequencies: whole-number multiples of the fundamental frequency

Chapter 18 Photoelectric emission

photoelectric emission: emission of electrons from a surface when illuminated with electromagnetic radiation of sufficient frequency

THINGS YOU NEED TO KNOW

threshold frequency: minimum frequency that will cause photoelectric emission from a material

quantum: small packet of energy

photon: smallest amount of light you can get at a given frequency

work function: minimum amount of energy needed to release an electron from the surface of a metal

Chapter 19 Einstein's photoelectric equation

stopping voltage (or **stopping potential**): voltage across a photocell that is just sufficient to stop photoelectrons reaching the receiving electrode

threshold frequency: the frequency below which no photoelectrons are emitted

Chapter 20 Photocurrents

electronvolt: energy transferred to an electron when it moves through a potential difference of 1 V; equivalent to 1.6×10^{-19} J

saturation: when all the emitted electrons are received by the other electrode and the maximum current flows

Chapter 21 Spectra

ground state: atom with its electrons in their lowest energy positions

excited: atom with one or more electrons raised above their ground-state positions

excitation energy: energy required to raise an electron to a position above its ground state

ionisation: when an electron is completely freed from an atom, leaving behind a positive ion

ionisation energy: energy required to free an electron from the ground state of an atom

emission spectrum: range of frequencies emitted by an atom when its electrons move between their allowable energy levels

Chapter 23 Matter waves

thermionic emission: process whereby electrons are emitted from a hot filament

Chapter 24 Star spectra

emission spectrum: a spectrum emitted by a hot gas which is characteristic of the elements it contains

absorption spectrum: a spectrum produced by light passing through a gas which has dark lines, where the gas has absorbed light, characteristic of the elements in the gas

Doppler effect: change of frequency of light caused by an observer moving relative to a light source

red-shift: an apparent decrease in frequency caused by the source and the observer separating

blue-shift: an apparent increase in frequency caused by the source and the observer approaching

Hubble's law: recession speed ∝ to distance of galaxy from us.

Chapter 25 The Big Bang and the Big Crunch

Big bang: a large and unexplained explosion that may have started the Universe

closed universe: our universe having sufficient mass for gravitational forces eventually to pull it back together

open universe: our universe not having enough mass so that gravitational forces are too weak to pull it back together

Equations to learn

Centripetal force	$F = \dfrac{mv^2}{r}$
Speed of waves	$c = f\lambda$

Index

Page references in **bold** refer to a definition/explanation in the 'Things you need to know' section

INDEX